# AGRO TEXTILES AND ITS APPLICATIONS

# AGRO TEXTILES AND ITS APPLICATIONS

Grace Annapoorani

**WOODHEAD PUBLISHING INDIA PVT LTD**

New Delhi

Published by Woodhead Publishing India Pvt. Ltd.
Woodhead Publishing India Pvt. Ltd.,
303, Vardaan House, 7/28, Ansari Road,
Daryaganj, New Delhi - 110002, India
www.woodheadpublishingindia.com

First published 2018, Woodhead Publishing India Pvt. Ltd.
© Woodhead Publishing India Pvt. Ltd., 2018
Reprint 2020

Woodhead Publishing India Pvt. Ltd. ISBN: 978-93-85059-36-0
Woodhead Publishing India Pvt. Ltd. e-ISBN: 978-93-85059-89-6

Typeset by Allen Smalley, Chennai

Printed and bound in India by Replika Press Pvt. Ltd.

# Contents

# Preface

Technical textiles offers new ways, means and opportunity to the Indian textile industry to sustain the present growth and thrive in near future. It would offer not only an opportunity to augment the growth, but also a new direction for advancement of the industry. The field of technical textiles had not received adequate importance in Indian context so far; however, it is a potential area where the textile industry can excel.

Traditional textiles today are unable to cope with cost of production for various reasons like technological obsolescence, high cost of modernization, power, etc. Present product mix of traditional textiles is not remunerative enough, and therefore, more and more ideas of value addition to textile products are gaining momentum. Technical textiles, in this context, are just perfect. Today, the term technical textiles has emerged as the most widely acceptable term for this expanding field of textile applications.

Agriculture and textile can play a duo by complementing the strengths of each other, to produce a new evolution of 'Agro textiles revolution'. It also gives multidimensional views and solutions to the problems being faced by agro industry. With the continuous increase in population worldwide, stress on agricultural crops has increased. Realizing the need of tomorrow, agricultural sector is opting for various technologies to get higher overall yield, quality and tasty agro products. Therefore, high-tech farming techniques, where textile structures are used, are being adopted to meet the purpose.

It is my privilege to author this book. This book was motivated by the inspiration gained during our literature survey in agro textiles. A prominent gap was felt to gather concise information in one place. The information available were scattered in some published papers, conference proceedings, websites and on some catalogues. The main aim of this book is to provide in-depth knowledge to all the readers, students, researchers, industrialists, academicians as well as professionals in agriculture and to all personnel involved in the production and manufacturing of agro textiles.

For clear understanding the text is supplemented with photographs, illustrations and tables. This book is designed for students studying textiles and fashion at higher and undergraduate level as well as those needing a comprehensive and authoritative overview of textile materials and process.

I am greatly indebted to my family members for their constant support during writing this book. I also thank my scholars for their tremendous technical support for this manuscript preparation. I acknowledge my special thanks to Woodhead Publishing India, for their support for making this book a reality. Above all I thank God Almighty for his abundant blessings. Hope this book will help the academicians and researchers in the field of agro textiles in acquiring information on the subject matter.

**Dr. S. Grace Annapoorani**

# 1

# Introduction

## 1.1 Introduction to Technical Textiles

Technical textiles are generally recognized to be one of the most dynamic and promising areas for the future of the textiles industry. Technical textiles are textile material and products manufactured primarily for their performance and functional properties rather than aesthetic or decorative purpose. Aesthetic properties are not much important for the technical textiles.

Although 'technical' textiles have attracted considerable attention, the use of fibres, yarns and fabrics for applications other than clothing and furnishing is neither a new phenomenon nor is exclusively linked to the emergence of modern artificial fibres and textiles. Natural fibres such as cotton, flax, jute and sisal have been used for centuries (and still are used) in applications ranging from tents and tarpaulins to ropes, sailcloth and sacking. There is evidence of woven fabrics and meshes being used in Roman times and before to stabilize marshy ground for road building – early examples of what would now be termed geotextiles and geogrids.

In some of the most developed markets, technical products already account for as much as 50% of all textile manufacturing activity and output. The technical textile supply chain is a long and complex one, stretching from the manufacturers of polymers for technical fibres, coating and speciality membranes through to the converters and fabricators who incorporate technical textiles into finished products or use them as an essential part of their industrial operations. The economic scope and importance of technical textiles extend far beyond the textile industry itself and have an impact upon just about every sphere of human economic and social activity.

And yet this dynamic sector of the textile industry has not proved entirely immune to the effects of economic recession, product and market maturity and growing global competition which are all too well known in the more traditional sectors of clothing and furnishings. There are no easy paths to success, and manufacturers and converters still face the challenge of making economic returns commensurate with the risks involved in operating in new and complex markets. If anything, the constant need to develop fresh products and applications, invest in new processes and equipment and market to an increasingly diverse range of customers is more demanding and costly than ever. Technical textiles have never been a single coherent industry sector and market segment. It is developing in many different directions with varying speeds and levels of success. There is continual erosion of the barriers between traditional definitions of textiles and other 'flexible engineering' materials such as paper and plastics, films and membranes, metals, glass and ceramics. What most participants have in common are many of the basic textile skills of manipulating fibres, fabrics and finishing techniques as well as an understanding of how all these interact and perform in different combinations and environments. Beyond that much of the technology and expertise associated with the industry resides in an understanding of the needs and dynamics of many very different end-use and market sectors. It is here that the new dividing lines within the industry are emerging.

## 1.2      Technical Textiles or Industrial Textiles

For many years, the term 'industrial textiles' was widely used to encompass all textile products other than those intended for apparel, household and furnishing end-uses. This usage has seemed increasingly inappropriate in the face of developing applications of textiles for medical, hygiene, sporting, transportation, construction, agricultural and many other clearly non-industrial purposes. Industrial textiles are now more often viewed as a subgroup of a wider category of technical textiles, referring specifically to those textile products used in the course of manufacturing operations (such as filters, machine clothing, conveyor belts, abrasive substrates, etc.) or which are incorporated into other industrial products (such as electrical components and cables, flexible seals and diaphragms, or acoustic and thermal insulation for domestic and industrial appliances).

If this revised definition of industrial textiles is still far from satisfactory, then the problems of finding a coherent and universally acceptable description and classification of the scope of technical textiles are even greater. Several schemes have been proposed. For example, the leading international trade exhibition for technical textiles, Techtextil (organized biennially since the late 1980s by Messe Frankfurt in Germany and also in Osaka, Japan), defines 12 main application areas (of which textiles for industrial applications represent only one group):

- Agrotech: agriculture, aquaculture, horticulture and forestry
- Buildtech: building and construction
- Clothtech: technical components of footwear and clothing
- Geotech: geotextiles and civil engineering
- Hometech: technical components of furniture, household textiles and floor coverings

- Indutech: filtration, conveying, cleaning and other industrial uses
- Medtech: hygiene and medical
- Mobiltech: automobiles, shipping, railways and aerospace
- Oekotech: environmental protection
- Packtech: packaging
- Protech: personal and property protection
- Sporttech: sport and leisure

The search for an all embracing term to describe these textiles is not confined to the words 'technical' and 'industrial'. Terms such as performance textiles, functional textiles, engineered textiles and high-tech textiles are also used in various contexts, sometimes with a relatively specific meaning (performance textiles are frequently used to describe the fabrics used in activity clothing), but more often with little or no precise significance.

## 1.3    Classification of Technical Textiles

Technical textiles can be divided into many categories, depending on their end-use. The classification developed by Techtextil, Messe Frankfurt is widely used. The classifications and its applications are shown in Fig. 1.

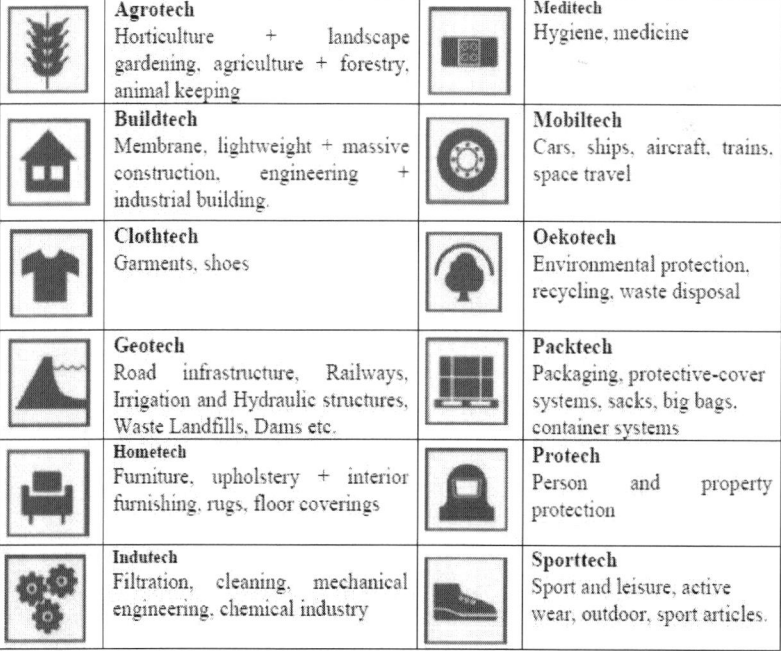

| | | | |
|---|---|---|---|
| | **Agrotech**<br>Horticulture + landscape gardening, agriculture + forestry, animal keeping | | **Meditech**<br>Hygiene, medicine |
| | **Buildtech**<br>Membrane, lightweight + massive construction, engineering + industrial building. | | **Mobiltech**<br>Cars, ships, aircraft, trains, space travel |
| | **Clothtech**<br>Garments, shoes | | **Oekotech**<br>Environmental protection, recycling, waste disposal |
| | **Geotech**<br>Road infrastructure, Railways, Irrigation and Hydraulic structures, Waste Landfills, Dams etc. | | **Packtech**<br>Packaging, protective-cover systems, sacks, big bags, container systems |
| | **Hometech**<br>Furniture, upholstery + interior furnishing, rugs, floor coverings | | **Protech**<br>Person and property protection |
| | **Indutech**<br>Filtration, cleaning, mechanical engineering, chemical industry | | **Sporttech**<br>Sport and leisure, active wear, outdoor, sport articles. |

Fig. 1   Classification of Technical Textiles

## 1.3.1    Agrotech (Agro textiles)

Textiles used in agriculture are termed as agro textiles. They are used for crop protection, fertilization, etc. The essential properties required are strength, elongation, stiffness and bio-degradation, resistance to sunlight and resistance to toxic environment. All these properties help with the growth and harvesting of crops and other foodstuffs. There is a growing interest in using materials which gradually degrade (biodegradables). Agro textiles are nowadays extensively being used in horticulture, farming and other agricultural activities. Agriculture and textiles are the largest industries in the world providing basic needs such as food and clothing. Agriculture is the largest industry in the world. Today, agriculture, horticulture area has realized the need of tomorrow and opting for various technologies to get higher overall yield, quality and tasty agro products. Adopting the hi-tech farming technique, where textile structures are used, could enhance quality and overall yield of agro products.

Textile structures in various forms are used in shade house/poly house, green house and also in open fields to control environmental factors like temperature, water and humidity. It also avoids agro products damage from wind, rain and birds. Agro textiles like sunscreen, bird net windshield, mulch mat, hail protection net, harvesting net, etc., can be used for achieving the above goal. Crop protection and weed control are the major challenges faced by the farmers in the agriculture industry. Textile industry is the second largest industry next to agriculture. Agro textiles contribute 8% share in the technical textiles break up. The usage of agro textiles will be benefited in terms of products with enhanced quality, higher yields fewer damages and bearable losses. The usage of agro textiles will be benefited in terms of products with enhanced quality, higher yields fewer damages and bearable losses. It also permits the usage of lower quantities of weed killers and pesticides. Agriculture has been among the most primal occupations of the humankind and is still a major industry, globally. In this era of modernization and high technological advancements, it has spread its horizons and started using man-made, non-conventional textiles, called 'technical textiles'.

## 1.3.2    Buildtech (Construction Textiles)

Textile materials are used in concrete reinforcement, façade foundation systems, interior construction, insulations, proofing materials, air conditioning, noise prevention, visual protection, protection against the sun and building safety. An interesting and aesthetic appealing application is the use of textile membranes for roof construction, which is also referred to as textile architecture. Polyvinyl chloride (PVC)-coated high tenacity polyethersulfone (PES), teflon-coated glass fibre fabrics or silicone-coated PES are used for their low creep properties. Splendid examples of such construction are found in football stadia, airports and hotels.

Civil engineering and building industry are an integral part of the development of human society as they involve the planning, design, building, operation and maintenance of infrastructure. The venturing of technical textiles or high-performance textiles in this sector has given a great impetus to the quality of construction. These textiles are used in the construction of buildings, dams, bridges, tunnels and roads and collectively comprise the 'Buildtech' sector. They offer mechanical properties such as lightness, strength and resilience as well as resistance to many factors such as creep, degradation by chemicals and pollutants in the air or rain and other construction material as well as the effects of sunlight and acid. These textiles play an important role in the modernization of infrastructure.

### 1.3.3     Clothtech (Clothing Textiles)

These are the clothing textiles, also known as Clothtech, including all the textile products that represent functional, most often hidden, components of clothing and footwear such as interlinings, sewing thread, insulating fibrefill and waddings. They are the 'high performance' garment fabrics whose demand is increasingly rising with the time.

The skin is the principal element that separates and protects the human body from the environment around it. It is also acts as a major exchange system of energy (e.g., heat) and matter (fluids and gases such as water, oxygen, etc.) between body and environment. Clothing as an artificial second skin has always been used by humans to enhance the protective function of their own skin. However, such additional protection often has a negative effect upon the exchange functionality of the human skin, in certain cases very severely like in the case of full body armour, firefighters, uniforms or diving suits. Functional and smart or intelligent clothing are the innovative response to such limitations. Functional clothing refers to products in which one or several specific functionalities are emphasized like strong insulation, water or fire resistance, breathability, wear resistance, etc. Smart clothing takes (multi) functionality one step further as it refers to products that can offer their functions in a more adaptive way in response to stimuli from the environment or the wearer.

Smart garments can for instance:

- Adapt their insulation function according to temperature changes.
- Detect vital signals of the wearer's body.

- Change colour or emit light upon defined stimuli.
- Generate or accumulate electric energy to power medical and other electronic devices.

Clothing is used to cover the body, to make you feel more attractive and to communicate with others. People wear clothes for many different reasons. Some of these reasons are physical. You wear clothes for comfort and protection. Others are for psychological and social reasons. Clothes give you self-confidence and express your personality. Clothes also help you identify with other people. All people have basic human needs. Meeting these needs provides satisfaction and enjoyment in life. Clothing helps to meet some of these needs. Knowing something about the role of clothing helps you to understand yourself and others better. Clothing is a complex but fascinating part of everyone's life.

## 1.3.4    Geotech (Geotextiles)

Geotextiles were one of the first textile products in human history. Excavations of ancient Egyptian sites show the use of mats made of grass and linen. Geotextiles were used in roadway construction in the days of the Pharaohs to stabilize roadways and their edges. These early geotextiles were made of natural fibres, fabrics or vegetation mixed with soil to improve road quality, particularly when roads were made on unstable soil. Only recently geotextiles have been used and evaluated for modern road construction. Geotextiles today are highly developed products that must comply with numerous standards. To produce tailor-made industrial fabrics, appropriate machinery is needed. Geotextiles have been used very successfully in road construction for over 30 years. Their primary function is to separate the sub base from the sub grade resulting in stronger road construction. The geotextile performs this function by providing a dense mass of fibres at the interface of the two layers. Geotextiles have proved to be among the most versatile and cost-effective ground modification materials. Their use has expanded rapidly into nearly all areas of civil, geotechnical, environmental, coastal and hydraulic engineering. They form the major component of the field of geosynthetics, the others being geogrids, geomembranes and geocomposites. The ASTM (1994) defines geotextiles as permeable textile materials used in contact with soil, rock, earth or any other geotechnical-related material as an integral part of civil engineering project, structure or system.

Geotextiles should fulfil certain requirements like it must permit material exchange between air and soil without which plant growth is impossible, it must be penetrable by roots, etc., and it must allow rainwater to penetrate the soil from outside and also excess water to drain out of the earth without erosion of the soil. To obtain all these properties in geotextiles, the proper choice of textile fibre is of paramount importance. The different synthetic fibres used in geotextiles are nylon, polyester and polypropylene while some natural fibres like ramie, jute, etc., can also be used. These are used in reinforcement of embankments or in constructional work. The fabrics in geotextiles are permeable fabrics and are used with soils having ability to separate, filter, protect or drain. The application areas include civil engineering, earth and road construction, dam engineering, soil sealing and drainage systems. The fabric used must have good strength, durability, low moisture absorption and thickness. Mostly non-woven and woven fabrics are used. Synthetic fibres like glass, polypropylene and acrylic fibres are used to prevent cracking of the concrete, plastic and other building materials. Polypropylene and polyester are used in geotextiles and dry/liquid filtration due to their compatibility.

## 1.3.5 Hometech (Domestic Textiles)

Hometech includes textiles used in a domestic environment such as interior decoration and furniture, carpeting, protection against the sun, cushion materials, fireproofing, floor and wall coverings and textile reinforced structures/fittings. In the contract market such as for large area buildings, ships, caravans, busses, fire retardant materials are used. Fire retardant properties are obtained either through the use of inherent fire retardant fibres such as modacryl or through the application of a coating with fire retardant additives (bromide of phosphorus compounds). The technical textiles are generating products (by combining the latest developments in advanced flexible materials with advances in process technologies) that ultimately affecting all sorts of consumer textile market, including both clothing and furnishings, these are called 'Hometech' products. 'The home textile market is recognized as an important part of the technical textile' that comprises household textiles, furnishings and upholstered furniture industry (including fibrefill and wadding applications in bedding, cushions, sleeping bags and furniture backings).

Home textiles transform a house into home by improving designs, patterns, size and styles. Today, people want modern and well-furnished homes in place of traditional and dull-looking house. Therefore, they are ready to accept new and expensive home textile products that could fill colours of newness and excitement to their life. Every furnishing in our house is inspired with latest designs such as wallpapers, upholstery, curtains, shower curtains, bedspreads and aprons. There are so many prints, shades and fabrics available in the market that make easy to pick one, which decorates the floor of the living hall, tiles of the kitchen, colour of the walls and wooden shade of the lobby. Hometech industry has been bright and striking place in the textile industry in the last decade in India. Designers, producers and the marketers endeavour to observe the consumer aspirations better. However, the Indian exports industry has its own troubles and concerns. In recent years due to consumer polarization, stiff competition within the industry and general access to media, role of textile design has changed in the country and elsewhere, both in terms of practice and appreciation. India has a strong traditional base in textiles which provides a unique place and resource that beautifully suits the consumer sensibility. The domestic Hometech market size, which is estimated at Rs. 600 billion in 2007–2008, is expected to touch Rs. 775 billion in 2010–2011. Exports are expected to grow at a compound annual growth rate of 12.3% from Rs. 120 billion in 2007–2008 to Rs. 170 billion in 2010–2011. The ability to produce and supply small quantities is not only limited to traditional handloom and handicrafts sector but it has now been demonstrated by mills and decentralized power looms as well. The underlying difficulty has been that we have yet not confidently acknowledged these strengths and built on that for the benefit of overall image of the industry. However, in recent years, a few progressive companies have demonstrated a remarkable business sense through this route.

## 1.3.6    Indutech (Industrial Textiles)

Textiles are used for chemical and electrical applications and textiles related to mechanical engineering. Silk-screen printing, filtration, plasma screens, propulsion technology, lifting/conveying equipment, sound-proofing elements, melting processes, roller covers, grinding technology, insulations, seals, fuel cell, etc. Indutech includes technical textile products used in the manufacturing sector such as conveyor belts, drive belts, cigarette filter rods, decatizing cloth, bolting cloth, absorption glass mat (AGM), glass battery separators, ropes and cordages, composites, filtration products, industrial brushes, etc. Indian Indutech market is estimated at US$ 854.4 million (2009–2010), about 35% of which is contributed by composites and around 27% by the ropes and cordages. The market for Indutech products is expected to grow at a rate of 11% annually and reach a size of US$ 1173 million by 2012–2013. Fibreglass products (including AGM glass battery separators) are expected to lead the growth in demand of Indutech technical textiles in India.

## 1.3.7    Medtex (Medical Textiles)

Medical textiles or Medtech is one of the most important, continuously expanding and growing field in technical textiles. Medical textiles represent structures designed and accomplished for a medical application. The number of applications is diverse, ranging from a single thread suture to the complex composite structures for bone replacement and from the simple cleaning wipe to advanced barrier fabrics used in operating rooms. Textile materials and products, that have been engineered to meet particular needs, are suitable for any medical and surgical application where a combination of strength, flexibility and sometimes moisture and air permeability are required. The medical textile industries have diversified with new materials and innovative designs. Recently, application of textiles has started going beyond the usual wound care, incontinence pads, plasters, etc. Latest innovation, i.e., wide variety of woven, non-woven, knitted forms of textile increasingly finding their way into a variety of surgical procedures. As the healthcare industry is growing enormously in India, the demand for the medical textile is also on the rise. These are commonly used in bandages and sutures (stitching the wounds). Not all the textile fibres can be used here, because their performances depend upon interaction with the cells and different fluids produce by the body. Sutures and wound dressing use fibres like silk and other synthetic fibres. Hollow synthetic fibres are used with nano or very small particles that are used for the delivery of drugs to any specific part of the body to prevent over dosage. Cotton, silk, polyester and polyamide are also used in medical applications. Medical textiles also cover surgical gowns and drapes. There are two classes of materials: reusables and non-wovens. Reusable is either PES or PES–cotton woven materials or laminates. In addition, non-woven materials are used in the operating theatre. High performance non-wovens are usually laminated with a plastic foil to provide for sufficient barrier properties to reduce wound infection.

### 1.3.8 Mobiltech (Textiles Used in Transport)

These textiles are used in the construction of automobiles, railways, ships, aircraft and spacecraft. The Mobiltech products comprise truck covers (PVC-coated PES fabrics), car trunk coverings (often needle felts), seat covers (knitted materials), seat belts, non-wovens for cabin air filtration (also covered in Indutech), airbags, parachutes, boats (inflatable), air balloons, nylon tyre cord, automotive carpets, headliners, insulation felts, sun blinds, helmets, seat upholstery, airline disposables, aircraft upholstery, webbings for aircrafts, etc.

### 1.3.9 Oekotech or Ecotech (Environment Friendly Textiles)

Oekotech or Ecotech includes any textile product, which is produced in eco-friendly manner and processed under eco-friendly limits.

- Eco textile is also known as Ecotech, Okeotech and eco-friendly textiles.

- Oekotech or Ecotech segment refers to use of technical textiles in environmental engineering.
- The new type clothes that are made out of organic cotton, hemp, bamboo, recycled polyester or Tencel (made from wood pulp) are the more eco-friendly fabrics.

New applications for textiles in environmental protection applications include floor sealing, erosion protection, air cleaning, prevention of water pollution, water cleaning, waste treatment/recycling, depositing area construction, product extraction, domestic water sewerage plants etc.

## 1.3.10    Packtech (Packaging Textiles)

One of the important uses of textiles is the manufacturing of bags and sacks, traditionally from cotton, flax and jute but increasingly from polypropylene. Products covered under Packtech range from polymer-based bags used for industrial packing to jute-based sacks used for packaging food grains and packaging used for tea. This packaging (excluding jute) is also referred to as flexible packaging materials. The ability to re-use these containers in many applications in place of disposable bags and sacks is another driver for their wider use. Products include polyolefin woven sacks [excluding flexible intermediate bulk container (FIBC)], FIBC, leno bags, wrapping fabric, jute Hessian and sacks (including food grade jute bags), soft luggage products (TT component), tea bags (filter paper), etc.

The Indian Packtech segment is expected to grow at a rate of 22% to US$11,782 million by 2016–2017 as per estimates of the Working Group on Textiles and Jute Industry, Ministry of Textiles, Government of India.

## 1.3.11    Protech (Protective Textiles)

These textiles are used for the protection against heat and radiation for firefighter clothing, against molten metals for welders, for bullet proof jackets, etc. All these things are obtained by usage of technical textiles with high performance fibres. In bullet proof jackets, special fibre aramid is used which has high tenacity, high thermal resistance and low shrinkage. Glass fibre is also used in fire proof jackets due to its high strength, chemical and flame resistance. Protective clothing used by the astronauts is covered with special chemicals, including lead to protect them from the sun's heat. Apart from their suits that are made from special fibres, their airship is lined with special fabric. Especially the beginning of the 21st century has witnessed an extraordinary period of innovation in textile science and new materials have transformed the familiar functions of textile to the advance: fabrics can collect solar energy and emit light or heat, fabrics can change colour, textiles protect human beings against chemical, mechanical and electrical radiation, etc. While all clothing is protective to some degree, the concern of protective functional textiles is not with routine needs, such as clothing for warm or cold, rainwear or routine work clothing. Protective functional textiles focus on more sophisticated needs, protection in situations where hazards or risks are present that have the potential to be life threatening or damage to the person working in and around the hazard. Protective textiles is manufactured using traditional textile production technologies such as weaving, knitting and non-wovens and also by specialized techniques such as 3D weaving, spacer fabric knitting and braiding using natural and man-made fibres, and finishing technology. Today, a wide range of high-performance fibres is commercially available for technical and industrial applications. These types of fibres are used in protective wear developed for impact protection and in textile reinforcement products for different applications. Many of the 'high-tech' fibres, such as Kevlar®, Nomex® (DuPont) and Twaron® (Acordis) aramids, Spectra® (Allied) HDPE fibres, PBI, Kermel® (RhonePoulenc), P84® (Inspec), carbon impregnated fibres, aramid spunlace materials, fibreglass, even steel, copper and other metal fibres have applications in the protective clothing areas. Conventional materials such as nylon and polyester, cotton and wool are also used and provide satisfactory protection in certain applications depending on the hazard or exposure. When new fibres with unique properties are being developed, they are often considered for protective clothing.

## 1.3.12    Sporttech (Sports Textiles)

Health is state of complete physical, mental and social well being and not merely the absence of disease or infirmity. Regular physical activity has a positive impact on major health risk factors, such as high blood pressure, high cholesterol, obesity and stress. Physical activity for nations is a cost-effective method to improve public health across populations. Thus participation in sports activities has increased remarkably owing to health and physical fitness. Sportswear is not just used by athletes while performing but is becoming a major part of everyday clothing during morning walks, jogging, yoga, stretching exercises and daily fitness activities because of quality comfort of sports clothing. It has been reported that only 30% of the sportswear manufactured is utilized by active sports person.

Selection of fibres or fabrics for manufacturing active sportswear is one big factor influencing performance, efficiency, ensuring protection and physical comfort. Technological developments have lead sportswear to a state of virtual insanity. The sports textiles sector includes specialist apparel for specific sports each with its own particular functions. The performance fibres, yarns, fabrics and finishes developed for this specialist sector are increasingly transferring to the mass market in the high street. The performance requirements of many sports goods often demand widely different properties from their constituent fibres and fabrics, such as barrier to rain, snow, cold, heat and strength and at the same time these textiles must fulfil the consumer requirements of drape, comfort, fit and ease of movement.

Among the contributing factors responsible for successful marketing of functional sportswear has been made in the fibre and polymer sciences and production techniques for obtaining sophisticated fibre, yarns and fabrics. Textiles used in the sports and leisure industries have diverse applications ranging from artificial turf used in sports surfaces to advanced carbon fibre composites for racquet frames, fishing rods, golf clubs and cycle frames. Other highly visible uses are balloon fabrics, parachute and paraglide fabrics and sailcloth. Sportech products include sports composites, artificial turf, parachute fabrics, ballooning fabrics, sail cloth, sleeping bags, sport nets, sport shoes components, tents, swimwear, etc.

# References

1. Proceedings, International Symposium on 'System Intensification Towards Food and Environmental Security', organized by Crop and Weed Science Society (CWSS), BCKV, Kalyani, WB, India, on February 24–27, 2011.

2. Horrocks, A.R., Anand, S.C. 'Handbook of Technical Textiles'. The Textile Institute, Woodhead Publishing Limited, Abington, Cambridge, 2000.

3. Kumar, P. 'Agricultural Performance and Productivity'. Rawat Publications, Jaipur, 2001, pp. 77–89.

4. Pawar, P.P., et al., Profitability potential of different-sized rice mills, Indian Journal of Agriculture Economics, Vol. 58, No. 3, 2003, pp. 612–614.

5. http://www.centexbel.be

6. http://www.technicaltextilesinfrance.com

7. http://textilelearner.blogspot.com

8. India – Textile Sector – Technical Textiles, US Commercial Service – Mumbai, 2014.

9. http://www.indiantextilejournal.com/articles/FAdetails.asp?id=1999

10. http://www.indiantextilejournal.com/articles/FAdetails.asp?id=4709

11. http://www.indiantextilejournal.com/articles/FAdetails.asp?id=884

12. http://www.indiantextilejournal.com/articles/FAdetails.asp?id=4020

# 2

# Agro Textiles

## 2.1    Introduction of Agro Textiles

Food, clothing and shelter are the three basic needs of human beings, and for clothing man had been using textile fibres right from the 'old stone age'. The latest developments in textiles and their industrial uses have lead to the birth and development in technical textile. Technical textile goods are mostly manufactured for non-aesthetic purpose where the function is its criteria. This is a very vast and rapidly developing sector that supports many industries.

India has tremendous potential for production, consumption and export of technical textile. Agro textile contributes about 1.5% to the total production of technical textile goods in India, while the globally growing demand for agricultural products is expected to boost the need for agro-textile products.

Technical textiles can be classified into many fields such as horticulture, agriculture, mining, civil, water, chemical, apparel, sports, defence, manufacturing, automotive, medical, paper, furniture, etc. Technical textiles used for agricultural applications are called as agro textiles. In the era of modernizing agriculture for high production, boost to agriculture will not be possible without increasing involvement of agro textiles (Fig. 2.1).

**Fig. 2.1**   Agro Textiles

The word 'agro textiles' is now used to classify the woven, non-wovens and knitted fabrics applied for agricultural and horticultural uses. The practice of textiles is also now widening to safeguard the agro products such as plants, vegetables and fruits from weather, weed, birds, etc.

Agriculture is looked upon as an industry in many parts of the world so as to ensure extensive use of all the land available in the country irrespective of the environmental factors. Controlling the environmental factors offers plants and crop protection for specific target and ensures quality maintenance by elimination of the variations and hazards associated with the weather. It is possible to control temperature and humidity with varying degree of precision, avoid damage from wind and rain, regulate nutrients level to meet plant requirement by using green/poly house techniques. Textile structures are used as controlling factors in greenhouse as well as in fields. The main areas of textiles in agriculture include applications in farming, animal husbandry and horticulture. The volumes of special textiles that are manufactured for agricultural applications are increasing day by day not only worldwide but also in India. In addition, a wide variety of textile products that are designed for general industrial applications are used for agriculture in large quantities. These products include hoses, ropes, conveyor belts, tyres, composites, containers, pots, filters, tarpaulins, sacks, etc. With the apparent exception of protective clothing for insecticides, farmers usually make use of existing fabrics to fulfil their needs.

The most important requirements of textiles for agricultural applications are weather resistance, resistance to microorganism, stable construction and lightweight. Therefore, synthetic fibres are the best choice for agricultural products. The use of textiles in horticulture (fruits, vegetables, nurseries and flowers) is increasing more rapidly than any other area in agriculture. Textiles in different forms are exclusively used for many agricultural end–uses, which includes knitted, woven, non-woven, extruded sheet, moulded product, ropes, belt, etc.

Man-made fibres gives advantages over natural fibres, mainly due to their favourable price/performance ratio, ease of transport as well as setting up, space saving storage and long service life. The use of non-woven, especially spun bounded, fabrics are increasing in agricultural applications at the expense of woven fabrics.

In the present era of globalization, the importance of agriculture as a prime mover particularly in developing countries has become unequivocal. Though agriculture is the backbone of India, textile can be the backbone of agriculture. The term 'agro textiles' is now used to encompass the woven, knitted and non-woven textiles used for agriculture, forestry, horticulture and landscape gardening, as well as in fishing and fish farming.

Agriculture, forestry, horticulture, floriculture, fishing segments, landscape gardening, animal husbandry, aquaculture and agro engineering sectors combined together are popularly called as Agrotech sector. Agro textiles are the application of textile materials in the above-mentioned sectors. With the continuous increase in population worldwide, stress on agricultural crops has increased. So it is necessary to increase the yield and quality of agro products. But it is not possible to meet fully with the traditionally adopted ways of using pesticides and herbicides. Today, agriculture and horticulture has realized the need of tomorrow and opting for various technologies to get higher overall yield, quality and tasty agro products. Agriculture today has very close links to the world of technical textiles, where principal functions being crop protection and the storage of produce. Reducing the use of herbicides and pesticides to limit environmental pollution is another achievement of agro textiles, along with improved efficiency and productivity to the agriculture sector.

Some growers are familiar with technical textiles, and yet for others, they are new products that are gradually coming up to replace materials traditionally used for crop protection such as plastic. For a third group of potential users, however, the possibilities they offer are still completely new.

Manufacturers in the textile industry are busy with designing fabrics and demonstrating the qualitative superiority and price competitiveness of innovative materials and methods for protecting crops over traditional ones. They are seeking to feed a market that is typically global where the requirements can differ radically according to climate and local needs.

In the industrial countries, the area dedicated to crop growing is shrinking; hence, agro textiles can help boost productivity by artificially enhancing climatic conditions and plant development. The global end-use consumption of agro textiles will increase from 3.3% in 2000 to 3.9% by 2010, according to a David Rigby Associates' study.

Thermal screens, on the contrary, are generally used to maintain the temperature inside the greenhouse, whereas the use of insect screen is used to restrict the damages caused by the insects or pests to the plants inside the greenhouse. Every plant has its own individual optimum requirements. By providing the right balance with the correct choice of covering material or its combination thereof, the optimum climatic conditions are created inside the greenhouses under which the plant's productivity is maximized. This study will help the farmers as well as the manufacturers in selecting suitable covering materials that are used in greenhouse applications in agriculture and horticulture field. The textile materials mostly produce by synthetics in various decompositions, utilized in the mode of either woven or non-wovens. The practice of textiles is also now widen to safeguard the agro products such as plants, vegetables and fruits from weather, weed, birds, etc. Textiles always keep up its style of uniqueness by creating vast technological strides in all the fields slowly since evolution.

With the continuous increase in population worldwide, stress on agricultural crops has increased. To keep grains, vegetables and flowers, it is necessary to increase the yield and quality of agro products. Today, agriculture, horticulture area has realized the need of tomorrow and opting for various technologies to get higher overall yield, quality and tasty agro products. Adopting the hi-tech farming technique, where textile structures are used, could enhance quality and overall yield of agro products.

Coir is a biodegradable organic fibre and hardest among other natural fibres. It is much more advantageous in different application for agricultural textiles. Coir is used commercially for the manufacture of wide range of products for varies end-use applications.

Textile structures in various forms are used in shade house/poly house, greenhouse and also in open fields to control environmental factors such as temperature, water and humidity. It also avoids damaging of agro products from wind, rain and birds. Agro textiles like sunscreen, bird net windshield, mulch mat, hail protection net, harvesting net, etc., can be used for achieving the above goal.

Agro textiles can be used inside greenhouses as well as in the open air, to keep areas safe and tidy. Agro textiles improve plant growth and crops in the orchards. Used mainly in planted areas, they provide weed suppression and ground moisture conservation, while allowing roots to breathe and water, air and nutrients to permeate through. This reduces upkeep, maintains higher soil temperatures and promotes more rapid and even plant growth. It is favoured by many leading landscape architects for its unrivalled performance, quality and price. Apart from these applications, agro textiles are widely used in agriculture, industries, homes and many other areas.

## 2.2     History of Agro Textiles

Agriculture is the backbone of our country. Now it is saying that textile can be the backbone of agriculture. Textile fabrics have a long history of use in Agrotech sectors to protect, gather and store products. Between the 18th century and the end of the 19th century, agricultural development was occurred, which saw a massive and rapid increase in agricultural productivity and vast improvements in farm technology. From then, textiles have always been used extensively in the course of food production, most notably by the fishing industry in the form of nets, ropes and lines but also by agriculture and horticulture for a variety of covering, protection and containment applications.

However, modern textile materials are also opening up new applications. Lightweight spun-bonded fleeces are now used for shading, thermal insulation and weed suppression. Heavier non-woven, knitted and woven constructions are employed for wind and hail protection. Fibrillated and extruded nets are replacing traditional baler twine for wrapping modern circular bales. Capillary non-woven matting is used in horticulture to distribute moisture to growing plants. Seeds themselves can be incorporated into such matting along with any necessary nutrients and pesticides.

The bulk storage and transport of fertilizers and agricultural products is increasingly undertaken using woven polypropylene flexible intermediate bulk containers – big bags – in place of jute, paper or plastic sacks. Today, modern textile materials and constructions have helped to increase the strength, lightness and durability of traditional products, as well as open up completely new markets.

Agriculture has been among the most primal occupations of the humankind and is still a major industry, globally. In this era of modernization and high technological advancements, it has spread its horizons and started using man-made, non-conventional textiles, called 'technical textiles'.

Tapping the potential of technical textiles and putting their vital properties to an advantage, agriculture, horticulture, forestry and fishing segments (all the four sectors combined together are popularly called as 'Agrotech' sector) are increasingly using them for equipment development and other applications.

This textile sector comprises of all textiles that are used in growing, harvesting, protection and storage of either crops or animals. It includes diverse items such as fishing nets and fishing lines, ropes, shade fabrics, mulch mats, woven and non-woven covers for crops, bird protection nests, etc. These textiles are driving the sector profitably by improving the productivity and reducing the need for chemicals.

## 2.3    Uses of Agro Textiles

- Preventing erosion and paving way for afforestation
- In greenhouse cover and fishing nets
- For layer separation in fields
- In nets for plants, rootless plants and protecting grassy areas
- As sun screens (since they have adjustable screening) and wind shields
- As packing material and in bags for storing grass (that has been mowed)

- Controlling stretch in knitted nets
- Shade for basins
- Anti-bird nets
- Fabrics for sifting and separation, for the phases of enlargement of the larvae
- Materials for ground and plant water management at the time of scarcity and abundance of water

# References

1. Hira, M.A. Agro-textile Products & Their Usage. Sasmira, Mumbai.

2. Sankhe Manoj, R.S. Chitnis: Textile structures and their application in agriculture, The Indian Textile Journal, Vol. CXIII, No. 3, 2002.

3. Basu, S.K. Agricultural and horticultural applications of agro textiles, The Indian Textile Journal, Vol. 121, Issue 12, 2011, pp. 141–148.

4. Kachru, R.P. Agro-Processing Industries in India—Growth, Status and Prospects, Indian Council of Agricultural Research, New Delhi, 2005, pp. 115–116.

5. Kaul, G.L. Horticulture in India – production, marketing and processing, Indian Journal of Agricultural Economics, Vol. 52, No. 3, 1997, pp. 561–573.

6. Kumar, P. Agricultural Performance and Productivity. Rawat Publications, Jaipur, 2001, pp. 77–89.

7. textilelearner.blogspot.com.tr/2014/04/applications-of-agro-textiles.html

8. www.technicaltextile.net › Articles › Agro Textiles

9. www.technotexindia.in/agro-textiles.html

10. www.bch.in/agro-textiles.html

# 3
# Fibres used for Agro Textiles

Man-made (synthetic) fibres are preferred for agricultural product than the natural fibres due to their high strength, durability and other suitable properi ties of agricultural applications. On the contrary, natural-fibre-based agro textiles not only serve the specific purpose but also after some year degrade and act as natural fertilizers. Man-made fibres are preferred for agricultural products than the natural fibres mainly due to their favourable price–performance ratio, ease of transport, space saving storage and long service life. Among synthetic fibres, polyolefin fibres are extensively used apart from small quantities of nylon, polyester fibres, whereas jute, wool, coir, sisal, flax and hemp fibres are the representative of natural fibres. Due to their high strength, durability and other suitable properties of agricultural applications, synthetic fibres are widely used in agrotech sector. Though man-made fibres (like polyolefins) are preferred for agro textiles than the natural fibres mainly due to their favourable price–performance ratio, lightweight with high strength and long service life, but natural fibres can be used in agro textiles in some specific arena where characteristics like high moisture retention, wet strength and biodegradability are effectively exploited.

Agro textile is an application of textile materials in the agriculture field. With the continuous increase in population worldwide, stress on agricultural crops has increased. So it is necessary to increase the yield and quality of agro products. But it is not possible to meet fully with the traditionally adopted ways of using pesticides and herbicides. Today, agriculture and horticulture have realized the need of tomorrow and opting for various technologies to get higher overall yield, quality and tasty agro products. The textile materials mostly produce by synthetics in various decompositions, utilized in the mode of either woven or non-wovens. The practice of textiles is also now widen to safeguard the agro products like plants, vegetables and fruits from weather, weed, birds, etc. Agriculture can play a due by complementing the strengths of each other, to produce a new evolution of 'agro textiles' revolution. It also gives you multidimensional views and solutions to the problems being faced by agro industry, from the textile sector. This is a wake up call to give a new shape to the developing trends of this novel technology in the form of woven, knitted and non-wovens forms. Textiles always keep up its style of uniqueness by creating vast technological strides in all the fields slowly since evolution. As technologists, we have emphasized on scope and need of the hour of agro textiles. Fibres used in agro-tech sectors are as follows.

## 3.1    Cotton

### 3.1.1    Origin and History of Cotton

Cotton is obtained plant source and it is classified as a natural material as it is obtained from the seeds of cellulose seed fibre staple fibre measuring 10–65 mm in length and white to beige in colour in its natural state. It is composed basically of a substance called cellulose. As cotton occupies 50% of the consumption of fibres by weight in the world, it is called as the king of all fibres. Cotton is the fabric for every home and is the most widely produced of textile fabrics today. It has now been proved that India was the first country to manufacture cotton. Among the recent findings at Mohenjo-Daro, there were a few scrapes of cotton sticking to the side of a sliver vase. Cotton is the white downy covering of the seed grown in the pods. The cotton plant grown in the tropics needs a climate with 6 months of summer weather to blossom and produce pods. The cotton fibre is the shortest of all the textile fibres. Its length varies from 8/10 of an inch to 2 inches. Cotton with short length fibres is technically known as 'short staple'. The one with the long fibres is called 'long staple' and it is more used since it is used for making fine qualities of cloth. Long staple is especially suitable as it is easy to spin and produces a strong smooth yarn. It is also suitable for mercerization a finishing process used to improve the absorbency, strength and lustre of fibre. Cotton is a soft, fluffy staple fibre that grows in a ball (Fig. 3.1) or protective case, around the seeds of the cotton plants of the genus *Gossypium* in the family of Malvaceae. The fibre is almost pure cellulose. Under natural conditions, the cotton bolls will tend to increase the dispersal of the seeds.

**Fig. 3.1**   Cotton Boll

The plant is a shrub native to tropical and subtropical regions around the world, including the Americas, Africa and India. The greatest diversity of wild cotton species is found in Mexico, followed by Australia and Africa. Cotton was independently domesticated in the Old and New Worlds. The fibre is most often spun into yarn or thread and used to make a soft, breathable textile. The use of cotton for fabric is known till date to prehistoric times; fragments of cotton fabric dated from 5000 BC have been excavated in Mexico and between 6000 BC and 5000 BC in the Indus valley civilization. Although cultivated since antiquity, it was the invention of the cotton gin that lowered the cost of production that led to its widespread use, and it is the most widely used natural fibre cloth in clothing today. No one knows exactly how old cotton is. Scientists searching caves in Mexico found bits of cotton bolls and pieces of cotton cloth that proved to be at least 7000 years old. They also found that the cotton itself was much like that grown in America today.

In the Indus River Valley in Pakistan, cotton was being grown, spun and woven into cloth 3000 BC. At about the same time, natives of Egypt's Nile valley were making and wearing cotton clothing.

Arab merchants brought cotton cloth to Europe about 800 AD. When Columbus discovered America in 1492, he found cotton growing in the Bahama Islands. By 1500, cotton was known generally throughout the world. Cotton seed are believed to have been planted in Florida in 1556 and in Virginia in 1607. By 1616, colonists were growing cotton along the James River in Virginia. Cotton was first spun by machinery in England in 1730. The industrial revolution in England and the invention of the cotton gin in the United States paved the way for the important place cotton holds in the world today.

Eli Whitney, a native of Massachusetts, secured a patent on the cotton gin in 1793, though patent office records indicate that the first cotton gin may have been built by a machinist named Noah Homes 2 years before Whitney's patent was filed. The gin, short for engine, could do the work 10 times faster than by hand. The gin made it possible to supply large quantities of cotton fibre to the fast-growing textile industry. Within 10 years, the value of the US cotton crop rose from $150,000 to more than $8 million.

## 3.1.2  Structure of Cotton

Cotton, the seed hair of plants of the genus *Gossypium*, is the purest form of cellulose readily available in nature. It has many desirable fibre properties making it an important fibre for textile applications. Cotton is the most important of the raw materials for the textile industry. The cotton fibre is a single biological cell with a multilayer structure. The layers in the cell structure are, from the outside of the fibre to the inside, cuticle, primary wall, secondary wall and lumen. These layers are different structurally and chemically. The primary and secondary walls have different degrees of crystallinity, as well as different molecular chain orientations. The cuticle, composed of wax, proteins and pectins, is 2.5% of the fibre weight and is amorphous. The primary wall is 2.5% of the fibre weight, has a crystallinity index of 30% and is composed of cellulose. The secondary wall is 91.5% of the fibre weight, has a crystallinity index of 70% and is composed of cellulose. The lumen is composed of protoplasmic residues. Cotton fibres have a fibrillar structure. The whole cotton fibre contains 88–96.5% of cellulose, the rest are non-cellulosic polysaccharides constituting up to 10% of the total fibre weight.

### 3.1.3    Physical and Chemical Properties of Cotton Fibre

#### 3.1.3.1    Length

Staple length of cotton fibre varies from 10 to 80 mm.

#### 3.1.3.2    Fibre Fineness

Wall thickness of different cotton ranges from 3.2 to 10 micron (μ). Ribbon width ranges from 12 to 25 μ. Cotton fibre fineness is measured at middle portion of fibre because the tip, that is, the end of the fibre is slightly tapered.

#### 3.1.3.3    Fibre Surface

The surface of cotton fibre, seen at high magnification, is wrinkled and striated and its cross section is kidney shaped.

#### 3.1.3.4    Tensile Strength

Cotton is a moderately strong fibre, and tenacity is 3.0–5.0 g/den. The strength is affected greatly by moisture and by the test conditions such as rate of loading and length of fibre section tested. The long, fine cottons, such as Sea Island and Egyptian, yield the strongest yarns and materials. Long and fine fibres of good strength can thus be spun into finer yarns.

#### 3.1.3.5    Elongation

Cotton does not stretch easily. It has an elongation at break of 5–10%.

#### 3.1.3.6    Elastic Properties

Cotton is a relatively inelastic, rigid fibre. At 2% extension, it has an elastic recovery of 74%; at 5% extension, the elastic recovery is 45%.

### 3.1.3.7    Density

The density of the cotton fibre is 1.54 g/cm$^3$.

### 3.1.3.8    Effects of Moisture

Under standard atmospheric conditions, cotton takes up about 6–8% of moisture regain. At 100% humidity, cotton has an absorbency of 25–27%.

Each unhydroglucose unit in cellulose has three hydroxyl groups. On an average, one out of these three hydroxyl groups on each glucose residue is a site for moisture absorption. Amount of moisture in cotton depends on temperature and relative humidity (RH).

The tensile properties of cotton fibres and yarns are affected appreciably by the amount of moisture absorbed by the fibres. Cotton yarns will continue to become stronger at high RH. Up to a RH of 100%, absorption of water by the cotton cellulose results in an increase in fibre strength. At higher humidity, fibre absorbs more water due to breakage of hydrogen bonds in amorphous region and availability of more hydroxyl groups. In addition, fibres saturated with water are about 20% stronger than dry fibres. The swelling of cotton yarns and fabrics in water is accompanied by some shrinkage.

### 3.1.3.9    Effect of Heat

Cotton has an excellent resistance to degradation by heat. It begins to turn yellow after several hours at 120°C and decomposes markedly at 150°C, as a result of oxidation. Cotton is severely damaged after a few minutes at 240°C. Cotton ignites easily and burns readily in air with odour similar to that of burning paper.

### 3.1.3.10    Effect of Aging

Cotton shows only a small loss of strength when stored carefully. It can be kept in the warehouse for long periods without any significant deterioration. After 50 years of storage, cotton may differ only slightly from fibre a year or two old. Ancient samples of cotton fabric taken from tombs more than 500 years old had four-fifths of the strength of new material.

### 3.1.3.11    Effect of Sunlight

There is a gradual loss of strength when cotton is exposed to sunlight and turns yellow. The degradation of cotton by oxidation when heated is promoted and encouraged by sunlight. It is particularly severe at high temperatures and in the presence of moisture. Much of the damage is caused by ultraviolet (UV) light and by the shorter waves of visible light. Under certain effects of weathering in direct sunlight can be serious. The cotton can be protected to some degree by using suitable dyes.

### 3.1.3.12    Effect of Acids

Cotton is attacked by hot dilute acids or cold concentrated acids, in which it disintegrates. It is not affected by cold weak acids.

### 3.1.3.13    Effect of Alkali

Cotton has an excellent resistance to alkalis. It swells in caustic alkali but is not damaged. It can be washed repeatedly in soap solutions without harm.

Cotton can be mercerized by treating with or without tension in a strong solution of alkali like sodium hydroxide (NaOH). Due to the mercerization, the swelling of the fibre takes place. The proper can be improved like:

- Shrinkage in yarn or fabric due to swelling
- Improvement in lustre
- Improvement in tensile strength
- Improvement in dyeability and its uniformity
- Improvement in dimensional stability

### 3.1.3.14    Effect of Organic Solvents

There are very few solvents that will dissolve cotton completely. It has a high resistance to normal solvents but is dispersed by the copper complexes cuprammonium hydroxide and cupriethylene diamine, and by concentrated (70%) $H_2SO_4$.

### 3.1.3.15    Microorganisms

Cotton is attacked by fungi and bacteria. Mildews, for example, will feed on cotton because of this rotting (decomposing) and weakening the material takes place. They have a characteristic musty smell and stain the fabric with naturally produced pigments. Mildews are particularly troublesome on cotton that has been treated with starchy finishes, and much of the damage can be avoided by thorough scouring. The pure cellulose is a less attractive food for mildew than the starch. Mildews and bacteria will flourish on cotton under hot, moist conditions. When cotton fabrics are to be used under conditions favourable to attack by microorganisms, they can be protected by impregnation with certain types of chemical. Copper compounds, such as copper naphthenate, will destroy organisms that would otherwise attack the cotton cellulose.

## 3.1.4 Applications of Cotton

1. Poplins and voiles are made by using cotton.
2. Cotton is used in great quantity as a fabric for hot weather wear.
3. The absorbency of cotton makes it an excellent material for household fabrics such as sheets and towels.
4. Cotton is widely used in making rainwear fabrics. It can be woven tightly to keep out the driving wind arid rain, yet the fabric will allow perspiration to escape.
5. Ventile fabrics, for example, are close-woven cotton materials of this sort which are given additional water resistance by a chemical proofing.
6. Heavy cotton yarns and materials are used for tyre cords and marquees, tarpaulins and industrial fabrics of all descriptions.
7. Cotton can be blended with other fibres like polyester, rayon to manufacture fabric for different applications.

## 3.2 Jute

Jute is the most environment-friendly fibre starting from the seed to expired fibre, as the expired fibres can be recycled more than once. Jute is a natural fibre popularly known as the golden fibre. It is one of the cheapest and the strongest of all natural fibres and considered as fibre of the future. Jute is the second to cotton in world's production of textile fibres. It is not only a major textile fibre but also a raw material for non-traditional and value-added products. One such example is jute agro textiles.

The intrinsic properties of jute have made it possible to develop different types of agro textiles suitable for specific applications. Jute is a lingo-cellulosic natural biodegradable and eco-friendly bast fibre. It has many advantageous properties like high tensile strength, high moisture absorption and retention capacity, unique drapability, porous structure, thermal resistance, etc.

## 3.2.1 Origin and History of Jute

Jute is a natural fibre with golden and silky shine and hence called the golden fibre. It is the cheapest vegetable fibre procured from the bast or skin of the plant's stem and the second most important vegetable fibre after cotton, in terms of usage, global consumption, production and availability. It has high tensile strength, low extensibility, and ensures better breath ability of fabrics.

Jute fibre is 100% biodegradable and recyclable and thus environmentally friendly. It is one of the most versatile natural fibres that have been used in raw materials for packaging, textiles, non-textile, construction and agricultural sectors. It helps to make best quality industrial yarn, fabric, net and sacks (Fig. 3.2).

Jute is the raw material for one of the India's oldest industries. The first jute mill started production in Bengal in 1856. After more than 150 years, the jute industry is now challenged by competition from alternative materials, by the recession in the international markets and by low awareness among consumers of the versatile, eco-friendly nature of jute fabric itself. Yet this industry still provides a livelihood to more than 250,000 mill workers and more than 4 million farmers' families. It is a golden bond with the earth; its use is a statement about ecological awareness as it is a fully biodegradable and eco-friendly fibre. It comes from the earth, it helps the earth and once its life is done it merges back into the earth.

**Fig. 3.2**   Jute

## 3.2.2    Structure of Jute

Jute is one of the strongest natural fibres. The long staple fibre has high tensile strength and low extensibility. Its lustre determines quality; the more it shines, the better the quality. It also has some heat and fire resistance. Jute has biodegradable feature. Jute includes good insulating and antistatic properties, as well as having low thermal conductivity and moderate moisture regain. It includes acoustic insulating properties and manufacture with no skin irritations. Jute has the ability to be blended with other fibres, both synthetic and natural, and accepts cellulosic dye classes such as natural, basic, vat, sulphur, reactive and pigment dyes. Jute can also be blended with wool. By treating jute with caustic soda, crimp, softness, pliability and appearance is improved, aiding in its ability to be spun with wool. Liquid ammonia has a similar effect on jute, as well as the added characteristic of improving flame resistance when treated with flame proofing agents (Fig. 3.3).

1. Jute fibre is 100% biodegradable and recyclable and thus environmentally friendly.

2. Jute is a natural fibre with golden and silky shine and hence called the golden fibre.

3. Jute is the cheapest vegetable fibre procured from the bast or skin of the plant's stem.

4. It is the second most important vegetable fibre after cotton, in terms of usage, global consumption, production and availability.

5. It has high tensile strength, low extensibility and ensures better breathability of fabrics. Therefore, jute is very suitable in agricultural commodity bulk packaging.

6. It helps to make best quality industrial yarn, fabric, net and sacks. It is one of the most versatile natural fibres which has been used in raw materials for packaging, textiles, non-textile, construction and agricultural sectors. Bulking of yarn results in a reduced breaking tenacity and an increased breaking extensibility when blended as a ternary blend.

7. Unlike the fibre known as hemp, jute is not a form of Cannabis. Therefore, it can be much more easily distinguished from forms of Cannabis that produce a narcotic.

8. Jute stem has very high volume of cellulose that can be procured within 4–6 months, and hence it also can save the forest and meet cellulose and wood requirement of the world.

9. The best varieties of jute are Bangla Tosha – *Corchorus olitorius* (Golden shine) and Bangla White – *Corchorus capsularis* (Whitish Shine). Mesta or Kenaf (*Hibiscus cannabinus*) is another species with fibre similar to jute with medium quality.

10. Raw jute and jute goods are interpreted as Burlap, Industrial Hemp, and Kenaf in some parts of the world.

The best source of jute in the world is the Bengal Delta Plain, which is occupied by Bangladesh and India.

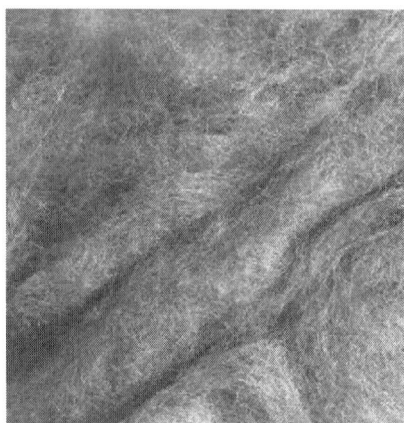

**Fig. 3.3**  Jute Fibre

### 3.2.3    Properties of Jute

Jute is a natural cellulosic fibre. It is the golden fibre of Bangladesh. Its important properties are discussed in the following sections.

#### 3.2.3.1    Length

The reeds of Jute fibre vary from 3 to 14 feet long, depending on the grade, and they show taper from root to end. Thick reeds contain coarse fibre and thin reeds contain finer fibre. It constitutes with ultimate fibres of average length of 2.5 mm.

#### 3.2.3.2    Fineness

It is a coarse fibre. Its diameter varies from 6 to 20 µ.

#### 3.2.3.3    Strength

Fibres are not so strong when compared with some other bast fibres but have good tensile strength. Fibres are naturally hard and brittle and break off with abrasion. Resistance to mechanical wear is low and not durable especially on exposure in moisture reduces its strength. Its extension at break is 2%.

#### 3.2.3.4    Colour

The best quality fibres are pale white or silvery grey, common qualities are brownish and greenish are inferior, roots are usually darker without any lustre. Better quality fibres shows matt and pitted surface with very poor strength.

*3.2.3.5    Lustre*

Better quality fibres have fairly high lustre but inferior quality fibres shows matt and pitted surface with very poor strength.

*3.2.3.6    Effect of Chemicals*

- Water: Jute is a hygroscopic fibre, that is, it takes in or gives out moisture to its surrounding atmosphere. Under standard testing atmosphere, moisture content value is 12.8% and moisture regain value of this fibre is 14.6%.
- Acid: This fibre is damaged by the action of strong acid hence wet processing on jute fibre is not done in acid medium.
- Alkali: It is safe in alkali medium; hence wet treatment is done on alkali medium.

*3.2.3.7    Effect of Biological Agents and Light*

It is attacked and damaged by the action of microbiological agents like bacteria, fungus, moths, insects, etc., in worm damp condition. Yellowing of the fibre is observed due to the effect of sunlight.

*3.2.3.8    Chemical Composite*

Chemical composite of jute fibre are as follows:
- Cellulose: 65.2%
- Hemi-cellulose: 22.2%
- Lignin: 10.8%
- Water: 1.5%
- Fats and wax: 0.3%

## 3.2.4    Applications of Jute Fibre

Jute is used for making yarn, twine, rope, sacking, cloth, hessian cloth, carpet backing cloth (CBC), carpet, mat, wall cloth, shopping bag and as packing materials. Jute is the second most important vegetable fibre after cotton; not only for cultivation, but also for various uses.

- Jute is used chiefly to make cloth for wrapping bales of raw cotton and to make sacks and coarse cloth.
- The fibres are also woven into curtains, chair coverings, carpets, area rugs, hessian cloth and backing for linoleum.

- While jute is being replaced by synthetic materials in many of these uses, some uses take advantage of jute's biodegradable nature, where synthetics would be unsuitable.
- Jute butts, the coarse ends of the plants, are used to make inexpensive cloth.
- Traditionally jute was used in traditional textile machineries as textile fibres having cellulose (vegetable fibre content) and lignin (wood fibre content). But, the major breakthrough came when the automobile, pulp and paper, and the furniture and bedding industries started to use jute and its allied fibres with their non-woven and composite technology to manufacture non-wovens, technical textiles and composites.
- Jute can be used to create a number of fabrics such as hessian cloth, sacking, scrim, CBC and canvas.
- Hessian, lighter than sacking, is used for bags, wrappers, wall coverings, upholstery and home furnishings.
- Sacking, a fabric made of heavy jute fibres, has its use in the name.
- Diversified jute products are becoming more and more valuable to the consumer today. Among these are espadrilles, floor coverings, home textiles, high performance technical textiles, geotextiles, composites, etc.
- Jute is also used in the making of ghillie suits which are used as camouflage and resemble grasses or brush.

## 3.2.5    Use of Jute in Agricultural Textiles

Jute agrotextile is a kind of natural technical textile, usually either in woven or non-woven form, made from 100% natural eco-friendly bast fibre of jute plant used on soil to achieve higher agricultural productivity by improving agronomical characteristics of soil and by reducing growth of unwanted vegetation.

The efficacy of jute agro textile has been established in the areas like:

- Soil conservation and reduction of nutrient loss.
- Nursery seed bed cover.
- Shade over nursery.
- Weed management and agro mulching.
- Afforestation in semi-arid zone.
- Sleeves for growth of sapling.
- Air layering and wrapping/covering of plants.

### 3.2.5.1    Mulch Mat

Mulch mats are used to suppress weed growth in horticulture applications. Weed control has traditionally been achieved with bark chips, jute or black plastic [polypropylene (PP)], which cover the soil, blocking out light and preventing the competitive weed growth around seedlings. Wool and coir non-woven fabrics are also effectively used as mulch mats (Fig. 3.4).

**Fig. 3.4**  Mulch Mats

### 3.2.5.2    Jute Sleeve for Nursery Use

For growth of seedling/sapling generally polythene bags (sleeve) are used in the nurseries. During transplantation of the grown up plants, the poly sleeves are torn off and thrown away on the ground causing pollution to the environment. Moreover, the experts opine that during tearing the polythene, there is chance of damaging of some important root network of the plants. In addition, air and water circulation in the soil mass inside the polythene are also affected. Due to its inherent nature, temperature inside the sleeve sometimes goes up in tropical areas and ice formation is observed in high altitude zone, causing high degree of mortality in young plants.

Jute-based woven as well as braided sapling bags are developed by Indian Jute Industries' Research Association to overcome the problems associated with poly sleeves. Due to the inherent properties of jute and the porous structure of the woven and braided sleeves, air and water circulation inside the sleeve are regulated, thus stimulating the plant growth (Fig. 3.5).

**Fig. 3.5**   Jute Sleeve for Nursery

### 3.2.5.3    Agro Bags

Jute has enjoyed a position of importance as packaging material since it had entered the packaging arena in 1793. Being a renewable and inexpensive natural fibre, abundantly available in Indian subcontinent, superior tensile strength and excellent frictional properties made jute the most widely accepted flexible packaging material. Agricultural commodities like grains, sugar, spices, vegetables, etc., are packed in jute bags domestically as well as globally since more than two centuries and have thus played a very important role in the national and international trade and commerce shown in Fig. 3.6.

**Fig. 3.6**   Jute Sleeve for Nursery

## 3.3    Wool

Wool is an animal hair fibre that, chemically, is made of a naturally occurring protein called keratin. This is a general definition that would include the body hairs of many animals but, in common parlance, the word is used for the body hair of sheep. This animal natural fibre comes from sheep fleece. The chemical structure is constituted by the keratin, a complex organic substance formed by amino acids. Its cross section is round, and its structure is scaly. Wool is composed of carbon, hydrogen, nitrogen and this is the only animal fibre, which contains sulphur in addition. The wool fibres have crimps or curls, which create pockets and give the wool a spongy feel and create insulation for the wearer. The outside surface of the fibre consists of a series of serrated scales, which overlap each other much like the scales of a fish. Wool is the only fibre with such serration's which make it possible for the fibres to cling together and produce felt (Fig. 3.7).

### 3.3.1    Origin and History of Wool

Like human civilization, the story of wool begins in Asia Minor during the Stone Age about 10,000 years ago. Primitive man living in the Mesopotamian Plain used sheep for three basic human needs: food, clothing and shelter. Later on man learned to spin and weave. As primitive as they must have been, woollens became part of the riches of Babylon. The warmth of wool clothing and the mobility of sheep allowed mankind to spread civilization far beyond the warm climate of Mesopotamia. Between 3000 and 1000 BC the Persians, Greeks and Romans distributed sheep and wool throughout Europe as they continued to improve breeds. The Romans took sheep everywhere as they built their Empire in what is now Spain, North Africa and on the British Isles. They established a wool plant in what is now Winchester, England, as early as 50 AD.

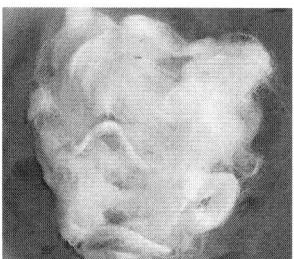

**Fig. 3.7**   Wool Fibre

### 3.3.2    Structure of Wool

Wool is the thick, wavy and fibrous protective covering of sheep. It mainly consists of the insoluble protein, keratin. The wool fibre grows from the follicle situated in the dermis (the middle layer of skin). It consists of three morphological components: the cuticle or skin, the cortex and medulla in the centre. True wool fibre (fine wool) does not contain the medulla (central core of hard cells) and has a hollow centre is shown in Fig. 3.8.

Fleece obtained from sheep is called grease wool or raw wool. Though wool fibres are more or less cylindrical, the surface consists of overlapping and interlocking scales of the cuticle. The serrated wool fibres tend to interlock and cling together imparting felting qualities to the wool. Wool fibre is elastic, hygroscopic, warmth retaining, durable, non-inflammable and transmits UV light.

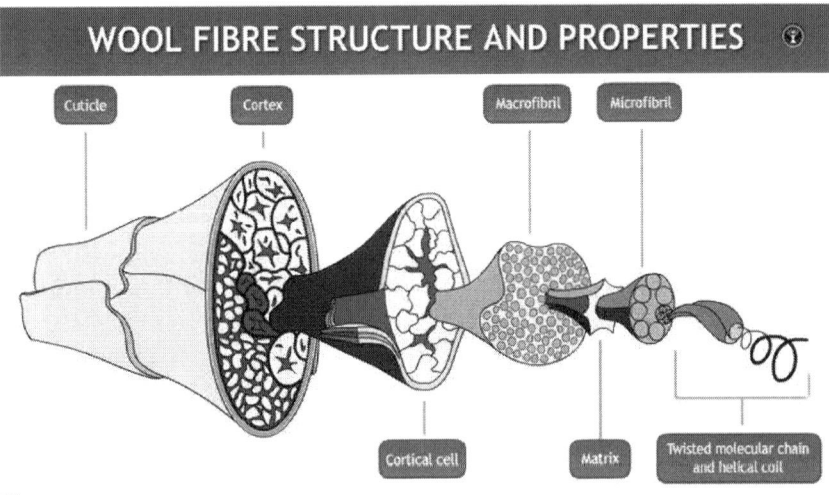

**Fig. 3.8**    Structure of Wool

### 3.3.3    Properties of Wool

The characteristics of wool fibre or protein fibres are as follows:

- They are composed of amino acids.
- They have excellent absorbency.
- Moisture regain is high.
- They tend to be warmer than others.
- They have poor resistance to alkalis but good resistance to acids.
- They have good elasticity and resiliency.

## 3.3.4    Applications of Wool

Wool fibres are traditional raw materials for textiles; according to the fibre diameter, they are designated to the clothing and interior textile industry. The end-products exploit the wool fibres' excellent mechanical and comfort properties. Technical textile is the field in which wool has made significant gains in recent years, building up on its various special characteristics. A major advantage of using wool for technical purposes is that the fibre diameter plays a minor role here, and this allows also cheap wool or even fibres from recycled textiles to be embedded into various products.

## 3.3.5    Use of Wool in Agricultural Textiles

Wool has better insulation properties under moist condition than PP/polyethylene (PE) and can prevent seedling damage from ground frost thus enabling earlier sowing and a longer growing season. The wool keeps the soil temperature constant and compared with black plastic, does not give a wind tunnel effect, which dries out the soil.

### 3.3.5.1    Mulch Mats

Needle-punched non-woven wool is used for mulch mats. The wool fibre biodegrades over 1–5 years and gets incorporated into the soil as fertilizer/conditioner for the next crop. Further, wool mulch mats allow water to enter into the soil (unlike black plastic), and also act as a barrier to prevent excessive soil desiccation during dry period.

Wool mulch mats permit water to enter into the soil and act as a blockade to put a stop to too much soil desiccation throughout dry period. It also offer better insulation and prevents damage from ground coolness.

## 3.4    Coir

Coir is a biodegradable organic fibre and hardest among other natural fibres. It is much more advantageous in different application for agricultural textiles. Coir is used commercially for the manufacture of wide range of products for various end-use applications.

Coir is a versatile hard fibre obtained from the husks of coconut. Coir is much more advantageous in different application for erosion control, reinforcement and stabilization of soil and is preferred to any other natural fibres. The fibre is hygroscopic in nature. Of all natural fibres, coir possess the greatest tearing strength, retained as such even in very wet condition.

### 3.4.1    Origin and History of Coir

Coir is the native name of the fibre extracted from coconut husk, the fibrous mass surrounding coconut, the fruit of the perennial plant cultivated extensively in the tropics. There is every reason to believe that the word 'Coir' has its origin from the Malayalam word 'Kayar' which means a cord, string or yarn spun out of fibre extracted from the husk of the coconut. Cocos fibre or coir extracted from the husk of the coconut is classed among the industrial hard fibres which enter the world market in the form of fibre, spun yarn or floor coverings.

Kerala, the land of coconut palm, got its name by this fact from the Sanskrit word 'Kera', the name for coconut tree. The word coir was supplied to the European vocabulary by Marco Polo, the Italian traveller.

### 3.4.2    Structure of Coir

Morphologically, coir is a multicellular fibre with 12–24 μ in diameter and the ratio of length to thickness is observed to be 35. Cells of the fibre surface are occasionally covered with stigmata. The chemical constituents have found to be cellulose, hemi-cellulose, lignin and pectin of which, cellulose and lignin are the major constituents (Fig. 3.9).

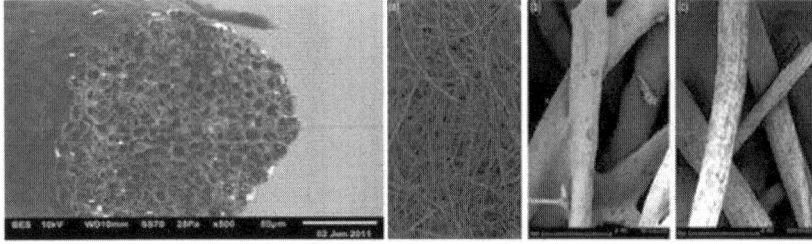

**Fig. 3.9**   Cross-sectional and Longitudinal View of Coir Fibre

### 3.4.3    Properties of Coir

Coco coir is a proven best alternative to any growing medium. Its use as a growing medium outperforms any other medium used for growing vegetables, ornamentals and tree plants. Its soft structure promotes easy root penetration and healthy growth. Coco coir is 100% environmentally friendly. It is a renewable resource that is consistent in quality. Coco coir has the best physical and chemical properties to promote better plant growth.

- Coco has high water-holding capacity. It can hold water up to eight times of its weight and release it over a period of time.

- Coco has ideal pH in the range of 6–6.7.
- It has excellent drainage and air porosity for better plant growth.
- Coco is very low in EC and carries mostly potassium salts, which is an essential major plant nutrient.
- Cation exchange capacity is very good.
- Coco coir has some antifungal properties that help plants to get rid of soil borne diseases. It inhibits pathogens like Pythium.
- Coco is very easy to re-hydrate after being dehydrated.
- It is a biodegradable source that degrades very slowly and has a life of 3–4 years.
- Contains significant amounts of phosphorous (10–50 ppm) and potassium (150–450 ppm).

As mentioned above, coco coir is not just a natural product with very good properties for plant growth, but it also has some winning advantages over other growing mediums. Some of the advantages are as follows:

- It is a 100% renewable resource.
- Coco coir is light in weight.
- It is consistent in high quality.
- Coco coir is completely environmentally friendly.
- The top of the product layer in grow bags/pots always remain dry, leaving behind no chances of fungal growth.
- Coco coir never shrinks, cracks or produces crust.
- It promotes better root systems in a short time.
- Coco coir is odourless, pleasant to handle and uniform in composition.

## 3.4.4    Applications of Coir

Coir being having the strong characteristics of retention of moisture is preferred for the agricultural applications. It is naturally resistant to rot, moulds and moisture. To suit specific applications, the coir fibre can be used as thus or by making a suitable product, which adapts the specific needs. Coir can be converted to coir yarn and then to woven mesh matting, which is used mainly controlling soil erosion and conditioning the soil. One more conversion of coir is to coir non-woven, which is also used for controlling erosion and conditioning the soil by more ground cover and soil retention.

### 3.4.5    Use of Coir in Agricultural Textiles

Non-woven coir is also used for basket liners, mulching mats, grow sticks, cultivation mats for plants, roof green application, portable lawn or instant lawn and many more applications. The coir fibre is also used for coco logs and coco beds for shore protection and stream banks.

#### 3.4.5.1    Erosion Control Blankets

The mesh of woven coir matting acts as miniature dams and prevents the seed or seedlings, which used to be washed away by rain and wind and facilitating the growth. The netting breaks up run off from heavy rains and dissipates the energy of flowing water. Once the growth of vegetation is occurred, the function of the coir is over and the vegetation will take over the protection of soil further.

Non-woven erosion blanket protects the soil from effective erosion and creating microclimates and mulching action.

#### 3.4.5.2    Mulch Blankets

Coir non-woven or closely woven matting acts as a filter allowing the water to flow across its plane as well as separator. The mulch mats will suppress the weeds and retain moisture in the soil, which will protect the roots from winter frosts.

#### 3.4.5.3    Coir Bed Blankets for Seed Germination

Coir plant bed blankets are ideal for germinating seeds and have been used in applications like wetland restoration, floating islands, aquatic spawning and even construction sediment traps. Their ability to hold a large amount of water means roots remain moist even in dry weather. As a 100% natural and biodegradable product, it offers a cost-effective and environmental-friendly solution.

#### 3.4.5.4    Basket Liners

Coir basket liners are used for hanging baskets. These coir pads facilitate better aeration of the growing media. Air can flow on more easily through the pores of coir pad; it will help the roots to grow faster and more vigorously. Coir non-woven felt cut in different shapes depending upon the size of the wire basket are used as basket liners.

#### 3.4.5.5    Bio-rolls

Coir non-woven felt mats made in the form of rolls filling it with peat moss/coir pith composite are used for bio-rolls. Rapid root growth is observed using these bio-rolls.

### 3.4.5.6    Roof Green Mats

Roof greening mats are manufactured with coir non-woven felt spread with seeds or seeds in laid with stitch bonded coir pads. These roof greening mats will spread on the roof surface and the seeds on the coir pads will sprout out and grow evenly on the surface.

### 3.4.5.7    Grow Sticks

Grow sticks are used as natural support for plant and creepers. They consist of wooden pole wrapped with the layer of coir fibre or non-woven felt. The roots of the plant can easily penetrate on the pores of coir pad.

## 3.5    Sisal

Sisal fibre is one of the most widely used natural fibre and is very easily cultivated. It is obtained from sisal plant. The plant is known formally as *Agave sisalana*. These plants produce rosettes of sword-shaped leaves which start out toothed, and gradually lose their teeth with maturity. Each leaf contains a number of long, straight fibres which can be removed in a process known as decortication. During decortication, the leaves are beaten to remove the pulp and plant material, leaving the tough fibres behind. The fibres can be spun into thread for twine and textile production, or pulped to make paper products (Fig. 3.10).

**Fig. 3.10**  Sisal Plant

Sisal fibre is fully biodegradable, and green composites are fabricated with soy protein resin modified with gelatine. Sisal fibre, modified soy protein resins, and composites are characterized for their mechanical and thermal properties. It is highly renewable resource of energy. Sisal fibre is exceptionally durable and a low maintenance with minimal wear and tear. Its fibre is too tough for textiles and fabrics. It is not suitable for a smooth wall finish and also not recommended for wet areas (Fig. 3.11).

**Fig. 3.11**   Sisal Fibre

The fine texture of sisal takes dyes easily and offers the largest range of dyed colours of all natural fibres. Zero pesticides or chemical fertilizers used in sisal agriculture. It is a stiff fibre traditionally used in making twine, rope and also dartboards. Sisal fibre is manufactured from the vascular tissue from the sisal plant (*A. sisalana*). It is used in automotive friction parts (brakes, clutches), where it imparts green strength to performs, and for enhancing texture in coatings application.

### 3.5.1     Origin and History of Sisal Fibre

Sisal is the coarsest of the 'hard' vegetable fibres. There are many varieties of the *Agave* plant throughout the tropical and sub-tropical world, especially in the Central American region, but the most important variety for fibre production on a commercial basis are *A. sisalana* (and its hybrids, the most common name) and *Agave fourcroydes* (better known as henequen).

The East African sisal plant was originated in the Yucatan in 1983. A little later, sisal bulbils sent from Kew Gardens were planted in Kenya.

After a difficult start, sisal production in East Africa prospered and by the 1960s Tanzania production alone totalled some 230,000 tons. Production in East Africa has contracted materially over the past three decades in response to the continuing movement in end products away from the low value agricultural twine market into considerably higher value more specialized end products, such as carpets, wire rope cores, dartboards, speciality pulps, plaster reinforcement and handicrafts.

Production in 2008 was approximately 23,000 tons per annum in Tanzania, plus some 7000 tons of Lake Sisal (not exported), 23,000 tons in Kenya and 8000-10,000 tons in Madagascar. There is also production in Southern China, unquantified, but estimated to be around 25,000 tons (for domestic consumption) and smaller quantities in Mozambique, Venezuela and Cuba.

In Mexico henequen production (largely in the Yucatan peninsular) has fallen from a peak of about 160,000 tons in the 1960s to about less than 5000 tons today, all of which is converted into product locally.

Both China and Mexico are now large importers of sisal fibre than growers. The first commercial plantings in Brazil were not made until the late 1930s and the first sisal fibre exports from there were made in 1948. It was not, however, until the 1960s that Brazilian production really accelerated and the first of many spinning mills, largely devoted to the manufacture of agricultural twines, were established. Today Brazil is the major world producer of sisal at some 50,000–60,000 tons from a high of 130,000 tons only 5 years ago.

### 3.5.2    Structure of Sisal Fibre

Sisal fibres used for textile processing are multicellular fibres. In cross section, the fibre bundles are built up of about 100–200 single cells which are bonded together by natural gums. Single cells consist of thick walls with a central lumen and shape of single cell is polygonic. The cross section of sisal fibres is neither circular nor fairly uniform in dimension. The lumen varies in size but is usually well defined. Longitudinally the fibre is straight and without crimp, and approximately cylindrical in appearance. There are many knots and stripes on the surface of the fibre, which confirms that the fibre bundle is composed of many single cells which are arranged in straight parallel lines.

### 3.5.3    Properties of Sisal Fibre

1. Sisal fibre is exceptionally durable with a low maintenance with minimal wear and tear.
2. It is recyclable.
3. Sisal fibres are obtained from the outer leaf skin, removing the inner pulp.
4. It is available as plaid, herringbone and twill.
5. Sisal fibres are anti-static, do not attract or trap dust particles and do not absorb moisture or water easily.
6. The fine texture takes dyes easily and offers the largest range of dyed colours of all natural fibres.

7. It exhibits good sound and impact absorbing properties.

8. Its leaves can be treated with natural borax for fire resistance properties.

### 3.5.4    Applications of Sisal Fibres

From ancient times sisal has been the leading material for agricultural twine because of its strength, durability, ability to stretch, affinity for certain dyestuffs and resistance to deterioration in saltwater.

1. Sisal is used commonly in the shipping industry for mooring small craft, lashing and handling cargo.

2. It is also surprisingly used as the fibre core of the steel wire cables of elevators, being used for lubrication and flexibility purposes. Traditionally sisal was the leading material for agricultural twine or baler twine. Although this has now been overtaken by PP.

3. It is used in automobile industry with fibre glass in composite materials.

4. Other products developed from sisal fibre include spa products, cat scratching posts, lumbar support belts, rugs, slippers, cloths and disc buffers.

5. Sisal is used by itself in carpets or in blends with wool and acrylic for a softer hand.

### 3.6    Flax

Flax is also called linen. The fibre is obtained from the stalk of a plant (*Linum usitatissimum*) which is from 80 to 120 cm high, with few branches and small flowers, of a colour which varies from white to intense blue, which flowers only for one day. Common flax was one of the first crops domesticated by man.

Common flax (also known as linseed) is a member of the Linaceae family which includes about 150 plant species widely distributed around the world. Some of them are grown in domestic flower beds, as flax is one of the few true blue flowers. Most 'blue' flowers are really a shade of purple.

### 3.6.1    Origin and History of Flax

Flax fibres are among the oldest fibre crops in the world and the use of flax for the production of linen goes back 5000 years. Pictures on tombs and temple walls at Thebes depict flowering flax plants. The use of flax fibre in the manufacturing of cloth in Northern Europe dates back to pre-Roman times. In the United States, flax was introduced by the Pilgrim fathers. Currently all flax produced in the United States and Canada are seed flax types for the production of linseed oil or flaxseeds for human nutrition. Flax fibre is soft, lustrous and flexible. It is stronger than cotton fibre but less elastic. The best grades are used for linen fabrics such as damasks, lace and sheeting. Coarser grades are used for the manufacturing of twine and rope. Flax fibre is also a raw material for the high quality paper industry for the use of printed currency notes and cigarette paper.

Flax is thought to have originated in the Mediterranean region of Europe; the Swiss Lake Dweller People of the Stone Age apparently produced flax utilizing the fibre as well as the seed. Linen cloth made from flax was used to wrap the mummies in the early Egyptian tombs. In the United States, the early colonists grew small fields of flax for home use, and commercial production of fibre flax began in 1753. However, with the invention of the cotton gin in 1793, flax production began to decline. Presently the major fibre flax producing countries are the Soviet Union, Poland and France.

## 3.6.2    Structure of Flax Fibres

**Fig. 3.12**    Flax Fibres

The flax is a thick, regular fibre with a subdued lusture. It ranges in length from about 10 to 100 cm, averaging about 50 cm. As flax fibres are strand of cells, its thickness depends upon the number of cells in one fibre cross section. Generally there are three to six cells present in a fibre cross section. The cells are about 25 mm long. The length to breadth ratio for flax fibre varies from 15,000:1 to 1500:1 for the long and short fibres, respectively (Fig. 3.12).

The colour of flax varies from light blonde to grey blond depending upon the agricultural and climatic conditions during growth and the quality of retting.

### 3.6.3    Properties of Flax Fibres

Flax is also known as Linen. Flax is a natural fibre and this fibre is used to make most of the expensive cloth which is most comfort to wear. Flax is much popular for the comfortness and softness.

Basically there are two types of properties of linen fibres. One is physical properties and another is chemical properties.

#### 3.6.3.1    *Physical Properties of Linen*

Physical properties of linen fibres are as follows:

1. Tensile strength: Linen is a strong fibre. It has a tenacity of 5.5–6.5 g/den. The strength is greater than cotton fibre.
2. Elongation at break: Linen does not stress easily. It has an elongation at break of 2.7 to 3.5%.
3. Colour: The colour of linen fibre is yellowish to grey.
4. Length: 18–30 inch in length.
5. Lusture: It is brighter than cotton fibre and it is slightly silky.
6. Elastic recovery: Linen fibre has not enough elastic recovery properties like cotton fibre.
7. Specific gravity: Specific gravity of linen fibre is 1.50.
8. Moisture regain (%): Standard moisture regain is 10–12%.
9. Resiliency: Very poor.
10. Effect of heat: Linen has an excellent resistance to degradation by heat. It is less affected than cotton fibre by the heat.
11. Effect of sunlight: Linen fibre is not affected by the sunlight as others fibre. It has enough ability to protect sunlight.

#### 3.6.3.2    *Chemical Properties of Linen*

Linen is a natural cellulosic fibre and it has some chemical properties. Chemical properties of the linen fibre are given below:

1. Effect of acids: Linen fibre is damaged by highly densified acids but low dense acids do not affect if it is wash instantly after application of acids.
2. Effects of alkalis: Linen has an excellent resistance to alkalis. It does not affected by the strong alkalis.

3. Effects of bleaching agents: Cool chlorine and hypochlorine bleaching agent does not affect the linen fibre properties.

4. Effect of organic solvent: Linen fibre has high resistance to normal cleaning solvents.

5. Effect of microorganism: Linen fibre is attacked by fungi and bacteria. Mildews will feed on linen fabric, rotting and weakling the materials. Mildews and bacteria will flourish on linen under hot and humid condition. They can be protected by impregnation with certain types of chemicals.

6. Effects of insects: Linen fibre does not attacked by moth-grubs or beetles.

7. Dyes: It is not suitable to dye. But it can be dye by direct and vat dyes.

## 3.6.4    Applications of Flax Fibres

Flax products have been used as sail and tent canvas, fishing lines, fishing nets, book binder's thread, leather working threads, sewing thread, suture thread, carpet warp and union cloth cotton and flax blended at weaving stag. Flax are also used to produce clothing, household, industrial and furnishing fabric only the best portion of seed flax can be use for wool-pile rugs backing, twine and rope. The linen fibre are extensively used in fine table damasks, handkerchiefs and sheer linen fabrics, linen and Dacron blends make excellent wash and wear fabric for dresses and sportswear.

Another secretes are waste flax fibres are make into high-grade bank notes, writing papers and cigarette papers. The linen makes from flax fibre is an excellent conductor of heat. Linen sheet are cold and linen garments are comfortable in hot weather.

## 3.7    Hemp Fibres

Hemp is the name of the soft, durable fibre that is cultivated from plants of *Cannabis* genus for industrial and commercial use. Hemp belongs to the mulberry family (Moraceae) and cultivated hemp varieties belong to the *Cannabis sativa* species. These hemp varieties can be very different in height and leafage. Hemp is one of the fibres which are usually named after their country of origin – thus Italian, Turkish, Chinese and Indian hemp varieties are available. It is an annual plant, grows in season from the middle of April to the middle of September. The plant can be monoecious or dioecious. The bast layer (phloem) of the stalk contains more important component of textile fibres. The fibre bundles form several layers in the bast and the bundles contain few unit cells.

The common application of the hemp fibre is to be blend with PP in a non-woven mat which through compression moulding technique turns to a three-dimensional part. When the hemp was compared with glass fibre has showed an equivalent Young's modulus, a much lower density and costs (approximately half the price), and a reduction in moulding time. Configurations of hemp fibre due to its properties, they can be used for automotive applications, sporting goods, musical instruments, luggage, etc., through the processes available like compression moulding, injection moulding and hand layup or even hybrid technologies.

Though man-made fibres (like polyolefins) are preferred for agro textiles than the natural fibres mainly due to their favourable price performance ratio, lightweight with high strength and long service life, but natural fibres can be used in agro textiles in some specific arena where characteristics like high moisture retention, wet strength and biodegradability are effectively exploited (Fig. 3.13).

**Fig. 3.13** Hemp Field

### 3.7.1     Origin and History of Hemp Fibres

Hemp is traditionally known as a fibre plant and most historical cultivation of the plant in the United States from the 17th to mid-20th centuries was with fibre use in mind.

Two kinds of fibres are derived from the hemp plant's stalk. These are long (bast) fibres and the short (core fibres). The long, strong bast fibres are similar in length to soft wood fibres and are very low in lignin content (lignin is the 'glue' that holds plants together). The short core fibres are more similar to hard wood fibres.

When grown as a fibre crop, hemp grows to a height of 6–12 feet without branching. Dense plantings (as many as 300 plants per square yard) help ensure that the plant grows straight. An ideal sized fibre plant has the same diameter as a #2 pencil (about ¼ inch or 6 mm). Male plants die after shedding pollen, but fibre crops are usually harvested before or during flowering. Hemp can be grown for dual use (seed and fibre harvest) but this practice has an impact on quality and quantity of fibre. A dedicated fibre crop yields the highest quality bast fibre for textiles and composites.

## 3.7.2    Structure of Hemp Fibres

Hemp is the bast fibre obtained from stems of *Cannabis sativa* L plants. It grows easily to a height of 4 m without agrochemicals and captures large quantities of carbon. The most important components of fibres are cellulose (77%), pectin (1.4%) and waxes (1.4%). Pectin is found in the middle lamellae and glues the elementary fibres to form bundles. The lignin (1.7%) is an incrusting component of the fibre. It is incrusting cellulose and contributes to the hardness and strength of fibres. It is located in the middle lamellae and fibre primary cell wall. Other components of hemp fibres are tannin, resins, fats, proteins etc. The content of these components is much higher in hemp than in cotton.

Hemp fibres are coarser when compared to flax and rather difficult to bleach. The fibres have an excellent moisture resistance and rot only very slowly in water. Hemp fibres have high tenacity (53-62 cN/tex); about 20% higher than flax, but low elongation at break (only 1.5%) is shown in Fig. 3.14.

**Fig. 3.14**   Hemp Fibres

## 3.7.3    Properties of Hemp Fibres

The fibre bundles are in several layers within the hemp stalks. A bundle consists of several fibres and bundles are connected by unit cells. The bundles in the inner layers are usually shorter and finer than those of the outer layers.

The shape of unit cells ranges from triangular to heptagonal with rounded corners and a large pith. Unit cells are connected by lignified pectins and the basis of hemp processing lies in loosening and dissolving this bond.

Its flexibility depends on the fineness of the bundle. The longer bundles require less twist during spinning. The elongation of the bundles is low and their flexibility is high and this can cause problems during spinning. Blending flax with hemp improves both the elongation and the flexibility of the yarns, which is low in 100% hemp yarns. However, these blends also decrease the strength of the yarns.

## 3.8    Polyethylene Fibres

The PE or polyethene [IUPAC name polyethene or poly (methylene)] is the most common plastic. The annual global production is around 80 million tonnes. Its primary use is in packaging (plastic bags, plastic films, geomembranes, containers including bottles, etc.). Many kinds of PE are known, with most having the chemical formula $(C_2H_4)_n$. PE is usually a mixture of similar polymers of ethylene with various values of $n$.

The PE has the simplest basic structure of any polymer. It has excellent electrical insulation properties over a wide range of frequencies, very good chemical resistance, good process ability, toughness and flexibility and transparency. It is a polymer of ethylene produced by addition polymerization. PE is a wax-like thermoplastic softening at about 80–130°C with a density less than that of water.

There are four quite distinct routes to the preparation of high polymers of ethylene: (a) high pressure process, (b) Ziegler processes, (c) the Phillips process and (d) the Standard Oil (Indiana) process.

### 3.8.1    Origin and History of Polyethylene Fibres

The PE is a polymer. Many number of ethylene monomers join with each in the synthesis of PE polymer. PE monofilaments were produced commercially on a small scale by conventional melt extrusion and drawing of polymers made by the high-pressure type of polymerization process, starting during World War II. PE is a hard, stiff, strong and a dimensionally stable material that absorbs very little water. It has good gas barrier properties and good chemical resistance against acids, greases and oils. It can be highly transparent and colourless but thicker sections are usually opaque and off-white. PE also has good self-extinguishing properties and resistance against UV. PE is obtained by the polymerization of ethane. Cationic coordination polymerization, anionic addition polymerization, radical polymerization and ion polymerization are the different methods by which PE can be produced. Every method gives different types of PE. Mechanical properties of PE depend on the molecular weight, crystal grouping and branching.

### 3.8.2    Structure of Polyethylene Fibres

The PE is a long chain aliphatic hydrocarbon and it is thermoplastic. $T_g$ is approximately $-120°C$. $T_m$ depends upon the structure which ranges from $108°C$ to $132°C$. It has high molecular weight alkaline and has a good resistance to chemical attack. Because it is a crystalline material and does not interact with any liquids, there is no solvent at room temperature. There are many grades of PE and differences arise from the following variables:

*   Variation in the degree of short chain branching in the polymer
*   Variation in the degree of long chain branching
*   Variation in the average molecular weight
*   Variation in the molecular weight distribution
*   The presence of small amount of co-monomer residues
*   The presence of impurities or polymerization residues some of which may be combining with polymer (Fig. 3.15).

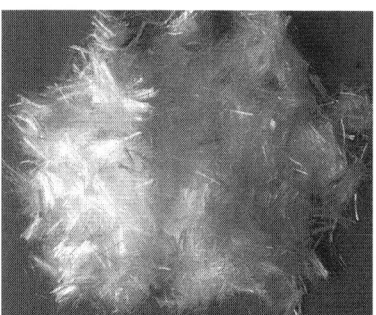

**Fig. 3.15**   Polyethylene Fibres

### 3.8.3    Properties of Polyethylene Fibres

The properties of PE can be divided into mechanical, thermal, chemical, electrical and optical properties.

#### 3.8.3.1    Mechanical Properties

The PE is of low strength, hardness and rigidity but has a high ductility and impact strength as well as low friction. It shows strong creep under persistent force, which can be reduced by addition of short fibres. It feels waxy when touched.

### 3.8.3.2    Thermal Properties
The usefulness of PE is limited by its softening point of 80°C (176°F) [high-density PE (HDPE), types of low crystalline softens earlier]. For common commercial grades of medium- and high-density PE, the melting point is typically in the range 120–180°C (248–356°F). The melting point for average, commercial, low-density PE is typically 105–115°C (221–239°F). These temperatures vary strongly with the type of PE.

### 3.8.3.3    Chemical Properties
The PE consists of non-polar, saturated, high molecular weight hydrocarbons. Therefore, its chemical behaviour is similar to paraffin. The individual macromolecules are not covalently linked. Because of their symmetric molecular structure, they tend to crystallize; overall PE is partially crystalline. Higher crystallinity increases density and mechanical and chemical stability.

### 3.8.3.4    Electrical Properties
The PE is a good electrical insulator. It offers good tracking resistance, however, it becomes easily electro statically charged (which can be reduced by additions of graphite, carbon black or antistatic agents).

### 3.8.3.5    Optical Properties
Depending on thermal history and film thickness, PE can vary between almost clear (transparent), milky-opaque (translucent) or opaque. Low-density PE thereby owns the largest, linear low-density PE slightly lower and HDPE the least transparency. Transparency is reduced by crystallites if they are larger than the wavelength of visible light.

## 3.8.4    Application of Polyethylene Fibres

1. Medical implants
2. Cable and marine ropes
3. Sail cloth
4. Composites like pressure vessel boat hulls, sports equipment, impact shields
5. Fish netting

6. Concrete reinforcement
7. Protective clothing
8. Can be used in radar protective cover because of its low dielectric constant
9. Can be used as a lining material of a pond which collects evaporation of water and containment from industrial plants
10. Useful in geotextile applications

## 3.9    Polypropylene

The PP, also known as polypropene, is a thermoplasticpolymer used in a wide variety of applications including packaging and labelling, textiles (e.g., ropes, thermal underwear and carpets), stationery, plastic parts and reusable containers of various types, laboratory equipment, loudspeakers, automotive components and polymer banknotes. An addition polymer made from the monomer propylene, it is rugged and unusually resistant to many chemical solvents, bases and acids.

The PP has a relatively slippery 'low energy surface' that means that many common glues will not form adequate joints. Joining of PP is often done using welding processes.

In 2013, the global market for PP was about 55 million tonnes. PP is the world's second-most widely produced synthetic plastic, after PE.

### 3.9.1    Origin and History of Polypropylene

The PP is the first stereoregular polymer to have achieved industrial significance. It is a 100% synthetic fibre. It is formed of 85% of monomer propylene. It is the by-product of petroleum. The fibres of PP were introduced to the textile market in the 1970. Now it has become fourth most important fibre classes after polyester, nylon and acrylic. Its use as apparel has been limited but it holds an important position in industrial applications. Structurally, it is a vinyl polymer and is similar to PE, due to the carbon atom in the backbone chain and a methyl group attached to it. It can be made from the monomer propylene by Ziegler–Natta polymerization and by metallocene catalysis polymerization.

### 3.9.2    Structure of Polypropylene Fibres

Initial polymerized PP was not crysatallizable and showed a low degree of polymerization and a low melting point which made it undesirable to use as fibre forming polymer. But with the work with Ziegler and Natta in the mid-1950s based on coordination catalyst system, now known as the Ziegler–Natta catalyst, it led to the preparation of a crystalline PP. This discovery resulted in stereoregular PP, with all methyl groups on one side, called isotactic PP. Isotactic PP is crystallizable with melting point of 160–174°C which could be easily converted into fibres capable of retaining molecular orientation at normal temperature (Fig. 3.16).

**Fig. 3.16** Polypropylene Fibres

### 3.9.3    Properties of Polypropylene Fibres

#### 3.9.3.1    *Molecular Weight*

Crystalinity of PP provides tensile strength by prevalent slippage of linear molecule. High molecular weight is necessary to obtain the good strength. Monofilaments with tenacities greater than 10 g/den require polymer of molecular weight greater than 100,000. In contrast, high-density PP fibres have molecular weight in the region 50,000 to 150,000, and low-density PP fibres have 20,000 to 25,000.

#### 3.9.3.2    *Mechanical Properties*

Medium tenacity of PP fibre is suitable for application. Commercial staple and continuous filament, the tenacity is in the range of 4.5–6 g/den. Tensile strength are 45,000–70,000 p.s.i. For use in ropes, nets high tenacity yarn are produced with tenacities up to 9 g/den. Special purpose yarn with tenacity up to 13 g/den can be made. Commercial used PP fibres have an elongation at break in the region of 15–25%. Multifilament yarns are in the range of 20–30% and staple fibre in 20–35%.

### 3.9.3.3    Thermal Property

The softening point of PP fibres is in the region of 150°C and the fibres melt at 160–170°C. The softening and melting point is determined by the way in which crystalinity has been influenced during treatment of fibre after spinning. PP fibre retains its flexibility to temperature of –70°C or lower. PP is thermoplastic hence can be heat set to desired shape and also is mouldable. Depending upon the type of PP fibre, it shrinks 5–12% at temperature above 100°C and softens at temperature about 140°C, and melts at temperature above 160°C and decomposes at temperature 288°C. The mechanical properties of the fibres deteriorate with increasing temperature below the softening point, but PP performs better than PE in this respect.

### 3.9.3.4    Chemical Properties

The PP is inert to a wide range of chemicals. Its resistance and susceptibility are similar to those of PE, but its high crystallinity tends to make it more resistant than PE to those chemicals which degrade olefin fibre. It shows good resistance against PP fibre.

## 3.9.4    Applications of Polypropylene

Because of its superior performance characteristics and comparatively low cost, PP fibre finds extensive use in the non-wovens industry. PP is a very important fibre in non-woven processing and dominates in many non-woven markets. The main application areas include: non-woven fabrics (Table 3.1), particularly absorbent product coverstock markets, home furnishings and automotive markets.

**Table 3.1**    Application of Polypropylene Fibres

| Application | Fibre Grade | Industry |
|---|---|---|
| Cigarette filter | Staple fibre 3 denier | Cigarette |
| Technical filters | Staple fibre 5 denier, needle punched non-woven | Wet filtration, excellent, chemical resistance, used in water, milk, bear, paints, coatings, petrochemicals, pharmaceuticals, filtration |
| PP woven socks | PP film fibre, with 10–15% LDPE to reduce fibrillation and cost | Fertilizers, flour, wheat, sugar, cement |
| Ropes and Twines | PP film and fibre | Agriculture |
| PP bale warp | Spun-bonded PP | Synthetic fibres |
| PP tapes | High modulus PP obtained by increasing draw ratio | Construction material like asphalt and concrete |
| PP construction / industry fabrics | Filling grade and staple fibre | Construction materials like asphalt and concrete |

| Substrate fabrics | Non-woven needle punched 3–4 denier staple fibres | Furniture fabrics as backing material for visual furniture fabrics, it serves as reinforcement. Also used for wall covering, luggage, table clothes, tarpaulins and automobile |
|---|---|---|
| Outdoor applications | Heavy deniers containing stabilizers, UV absorber, etc. | Sports |
| Non-electric fuses for initiating explosives | PP slit film tapes | Mining industry |
| Medical/surgical disposable fabric | PP staple fibre non-wovens, face masks | Hospital |

## 3.10    Nylon

Nylon was the first man-made synthetic fibre to be commercialized (1939). It is a polyamide fibre with amide linkage –NHCO–. It is derived from a diamine and a dicarboxylic acid. There are two methods to obtain commercially important nylons.

Firstly, polycondensation of difunctional monomers utilizing either amino acids or stoichiometric pairs of dicarboxylic acids and diamines. The resulting nylon is named on the basis of the number of carbon atoms separating the two acid groups and the two amines. Second process includes the ring opening polymerization of lactams.

There are several commercial nylons, such as nylon 6, 11, 12, 6/6, 6/10, 6/12 and so on. Out of these, the most widely used in the textile industry are nylon 66 (polyhexamethylene diamide) which is obtained by first process and nylon 6 (polycaprolactam) is obtained by the ring opening polymerization. Nylon is produced by melt spinning and valued for its lightweight, incredible tensile strength, durability and resistance to damage. It takes dye easily, making the fabric available in a wide array of colours for consumers. Textile materials composed of nylon tend to be light in weight because of the low density of nylon.

### 3.10.1    Origin and History of Nylon

Nylon is a manufactured fibre in which the fibre-forming substance is a long-chain synthetic polyamide in which less than 85% of the amide linkages are attached directly (–CO–NH–) to two aliphatic groups.

Nylon is a synthetic polymer, a plastic, invented on 28 February 1935 by Wallace Carothers at the E.I. du Pont de Nemours and Company of Wilmington, Delaware, USA. The material was announced in 1938 and the first nylon products; a nylon bristle toothbrush made with nylon yarn (went on sale on 24 February 1938) and more famously, women's stockings (went on sale on 15 May 1940). Nylon fibres are now used to make many synthetic fabrics, and solid nylon is used as an engineering material.

## 3.10.2    Structure of Nylon

Staple fibres of nylon are usually crimped in configuration similar to viscose. Nylon filaments are textured for polyamide filaments to be used for applications such as wearing apparel and carpets. This process entails mechanical distortions of the filaments and result in yarns characterized by a greater apparent volume and increased stretching properties. The textured yarns are either fine tex (1.7–2.2 tex) for apparel application, mainly stretch fabrics or heavy tex (110–140 tex) primarily for carpets.

The length of the staple fibre is comparable to cotton or wool, depending upon the end-use of fibre. The diameter of nylon filaments or staple fibres ranges from about 14 to 24 µm. The fibre length to breadth ratio usually exceeds 2000:1, which ensures that even the shorter nylon staple fibres can be satisfactorily spun into yarn.

The colour of fibre is slightly off-white. Most of the incident light upon nylon is reflected with considerable intensity from filaments or staple fibre's smooth and regular surface. This results in harsh and bright lustre. For subdued lustre of fibre delustering agent titanium dioxide is added to the spinning solution. Its presence makes the nylon appear white.

Like other man-made fibres, it lacks a discernible fibre microstructure and impurities, thus permitting some light to pass through the fibre, which makes it translucent is shown in Fig. 3.17.

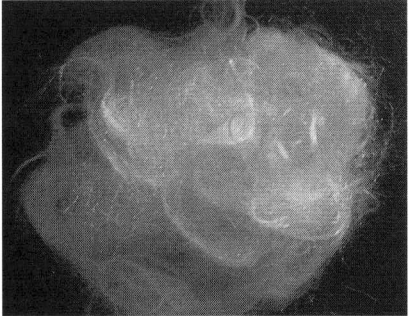

**Fig. 3.17**  Nylon

## 3.10.3    Properties of Nylon

### 3.10.3.1    Physical Properties

- Tenacity: 4–9 g/den (dry), in wet 90% of dry
- Elasticity: Breaking extension is 20–40%
- Stiffness: 20–40 g/den
- Moisture regain: 3.5–5%; (not absorbent due to crystallinity)
- Specific gravity: 1.14
- Abrasion resistance: Excellent
- Dimensional stability: Good
- Resiliency: Excellent
- Softening point: Nylon 6,6 – 229°C, nylon 6 – 149°C
- Melting point: Nylon 6,6 – 252°C, nylon 6 – 215°C
- Hand feel: Soft and smooth

### 3.10.3.2    Chemical Properties

- Acid: Nylon 6,6 is attacked by mineral acids is disintegrated or dissolved almost. But is inert to dilute acetate acid and formic acids even of the boil. It is dissolved in the concentrated formic acid. Nylon 6 is attacked by mineral acid but resistant to dilute boiling organic acid.
- Bleaches: Not attacked by oxidizing and reducing bleaches but may be harmed by chlorine and strong oxidizing bleaches.
- Alkali: Nylon is substantially inert to alkalis.
- Organic solvent: Most of the solvent have little or no effect on nylon. Phenol metacressol and formic acid dissolve the fibre but solvents used in stain removal and dry cleaning do not damage it.
- Light: No discolouration. Nylon 6 gradually loss of strength on prolonged extension.
- Biological: Neither microorganism nor moth, larvae attack nylon.
- Electrical: High insulating properties lead to static charges on the fibre.
- Flammability: Burns slowly.

## 3.10.4    Applications of Nylon

1. Nylon is a high strength fibre. It is used for making fishing nets, ropes, parachutes and type cords.
2. It is used for making fabrics in textile industry.
3. Crinkled nylon fibres are used for making elastic hosiery.
4. Nylon is widely used as plastic for making machine parts.
5. It is blended with wool to increase the strength.

## 3.11    Polyester

Polyester is a term used for 'long-chain polymers' chemically composed of at least 85% by weight of an ester which is organic salt formed from the reaction between an alcohol and an acid. Polyester also refers to the various polymers in which the backbones are formed by the esterification condensation of polyfunctional alcohols and acids. Polyester is a man-made, synthetic polymer, filament or staple fibre. Polyester textile filament or staple fibre is composed of PE terephthalate polymers.

Polyesters are a medium weight fibre with a density of 1.39 g/cm$^3$. As it is heavier than nylon, polyester textile materials are manufactured as 'thin' fabrics, since thick polyester fabrics are too heavy.

### 3.11.1    Origin and History of Polyester

Polyester began as a group of polymers in W.H. Carothers' laboratory. Carothers was working for duPont at the time when he discovered that alcohols and carboxyl acids could be successfully combined to form fibres. Polyester was put on the back burner, however, once Carothers discovered nylon. A group of British scientists – J.R. Whinfield, J.T. Dickson, W.K. Birtwhistle and C.G. Ritchie – took up Carothers' work in 1939. In 1941, they created the first polyester fibre called terylene. In 1946, duPont bought all legal rights from the Brits and came up with another polyester fibre which they named Dacron.

Polyester was first introduced to the American public in 1951. It was advertized as a miracle fibre that could be worn for 68 days straight without ironing and still look presentable.

In 1958, another polyester fibre called Kodel was developed by Eastman Chemical Products, Inc. The polyester market kept expanding. Since it was such an inexpensive and durable fibre, many small textile mills emerged all over the country – many located in old gas stations – to produce cheap polyester apparel items. Polyester experienced a constant growth until the 1970s when sales drastically declined due to the negative public image that emerged in the late 1960s as a result of the infamous polyester double-knit fabric!

Today, polyester is still widely regarded as a 'cheap, uncomfortable' fibre, but even now this image is slowly beginning to change with the emergence of polyester luxury fibres such as polyester microfibre.

### 3.11.2    Structure of Polyester

Polyester is fine, regular and translucent filament or staple fibre. Both filament and staple fibres are manufactured in crimped or textured configuration.

Crimping increases the inter-fibre friction, results in better fibre cohesion during and after spinning of its yarn and improved texture.

The length of the polyester filaments depends upon the size of the yarn package onto which it is wound. The length of the staple fibre is comparable to cotton or wool.

The diameter polyester filaments or staple fibres range from 12 to 25 μm. Diameter depends upon end-use requirements. The fibre length to breadth ratio usually exceeds 2000:1 and ensures that even the shorter polyester staple fibres can be satisfactorily spun into yarn.

The colour of fibre is slightly off-white. Most of the incident light upon polyester is reflected with considerable intensity from filaments or staple fibre's smooth and regular surface. This results in harsh and bright lustre. Like other man-made fibres, it lacks a discernible fibre microstructure and impurities, thus permitting some light to pass through the fibre, which makes them translucent (Fig. 3.18).

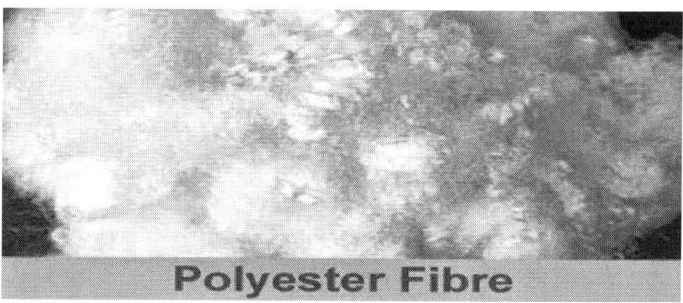

**Fig. 3.18**   Polyester

## 3.11.3    Properties of Polyester

### 3.11.3.1    Physical Properties

*Tenacity*: Polyester filaments and staple fibre are strong due to their crystalline nature. The crystalline nature permits for the formation of highly effective Van der Waals forces as well as since hydrogen bonds which provided the fibre its good tenacity. The tenacity remains unchanged when wet since the fibre resists the entry of water molecules to a significant extent.

*Elastic plastic nature*: The very crystallinity of the fibre prevents wrinkling and creasing. Repeated stretching and straining causes, distortion of the polymer system as the Van der Waals forces cannot withstand much stretching.

*Hygroscopic nature*: Filaments and staple fibres are hydrophobic. The lack of polarity and the very crystalline structure resists the entry of water molecules into the polymer system. The hydrophobic nature of the polymer system attracts fats, greases, oils, acid or any other greasy soils. It is believed to be oleophilic. The oleophilic nature makes it not easy to remove grease by soap but by dry-cleaning with organic solvents.

*Thermal properties*: It is a poor heat conductor and it has low resistance to heat. It melts on heating. Polyester textile materials can be permanently heat-set. It is a thermoplastic fibre meaning that it is capable of being shaped or turned when heated. Thermoplastic fibres heated under strictly controlled temperatures soften and can then be made to similar to a flat, creased or pleated configuration. When cooled, thermoplastic fibres retain the new configuration.

### 3.11.3.2    Chemical Properties of Polyester

*Effect of acids*: These polymers are resistant to acids.

*Effect of alkalis*: Alkaline conditions as seen in laundering hydrolyze the ester groups in polyester polymers. The crystalline nature prohibits hydrolysis to a greater extent and it is the surface of filament which gets hydrolyzed. Continued laundering results in hydrolysis and materials get fewer as the surface film of the fibre gets lost.

*Effect of bleaches*: It does not require bleaching. It retains its whiteness and requires only chlorine bleaches to be used when essential.

*Sunlight*: It withstands the sun's UV radiations and is resistant to acidic pollutants in atmosphere.

*Colour fastness*: It is not easy for dye molecule to penetrate the fibre when dyed, it retains its colour after regular wash.

*Microorganisms*: It is resistant to bacteria and other microorganisms.

## 3.11.4    Applications of Polyester

Polyester also has industrial uses as well, such as carpets, filters, synthetic artery replacements, ropes and films (Fig. 3.19).

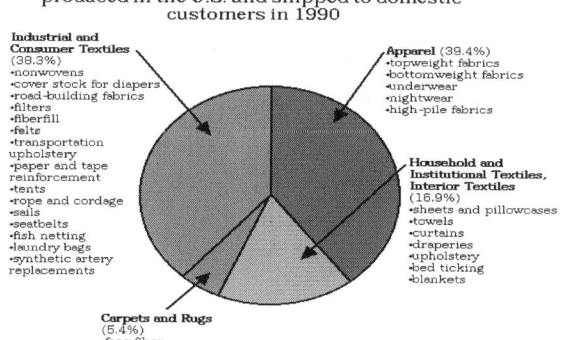

**Fig. 3.19**  Industrial Uses of Polyester

## 3.12    Polyolefin Fibres

Olefin fibre is a manufactured fibre in which the fibre-forming substance is any long-chain synthetic polymer composed of at least 85% by weight of ethylene, propylene or other olefin units. Olefin fibre is a generic description that covers thermoplastic fibres derived from olefins, predominately aliphatic hydrocarbons. Olefins are products of the polymerization of propylene and ethylene gases. PP and PE are the two most common members of the family. PP is extremely versatile as a fibre-forming material, whereas PE is not as good a fibre-forming high polymer material. Since its introduction into the textile industry in the 1950s, the list of successful products and markets for PP fibre has increased exponentially (Fig. 3.20).

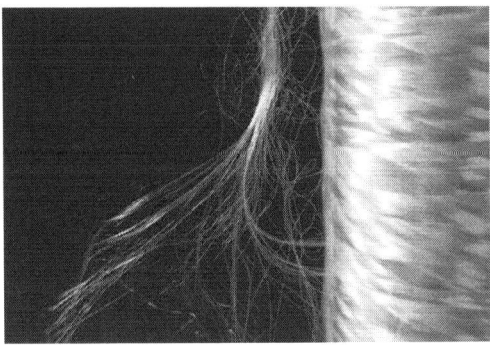

**Fig. 3.20**   Polyolefin Fibre

### 3.12.1    Origin and History of Polyolefin Fibres

Italy began production of olefin fibres in 1957. The chemist Giulio Natta successfully formulated olefin suitable for more textile applications. Both Natta and Karl Ziegler were later awarded the Nobel Prize for their work on transition metal catalysis of olefins to fibre, also known as Ziegler–Natta catalysis. The US production of olefin fibres began in 1960. Olefin fibres account for 16% of all manufactured fibres.

### 3.12.2    Structure of Polyolefin Fibres

*3.12.2.1    Physical*

Olefin fibres can be multi- or monofilament and staple, tow or film yarns. The fibres are colourless and round in cross section. This cross section can be modified for different end-uses. The physical characteristics are a waxy feel and colourless.

*3.12.2.2    Chemical*

There are two types of polymers that can be used in olefin fibres. The first, PE, is a simple linear structure with repeating units. These fibres are used mainly for ropes, twines and utility fabrics.

The second type, PP, is a three-dimensional structure with a backbone of carbon atoms. Methyl groups protrude from this backbone. Stereoselective polymerization orders these methyl groups to the same spatial placement. This creates a crystalline PP polymer. The fibres made with these polymers can be used in apparel, furnishing and industrial products.

### 3.12.3    Properties of Polyolefin Fibres

Olefin fibres have great bulk and cover while having low specific gravity. This means 'Warmth without the weight'. The fibres have low moisture absorption, but they can wick moisture and dry quickly. Olefin is abrasion, stain, sunlight, fire and chemical resistant. It does not dye well but has the advantage of being colourfast. Since Olefin has a low melting point, textiles can be thermally bonded. The fibres have the lowest static of all manufactured fibres and a medium lustre. One of the most important properties of olefin is its strength. It keeps its strength in wet or dry conditions and is very resilient. The fibre can be produced for strength of different properties.

### 3.12.4    Application of Polyolefin Fibres

In agro textiles, polyolefin is used for packaging of food materials.

- They have high specific strength, are lightweight, are resistant to water and waterborne microorganisms and are durable.
- They are compatible with food stuffs and prevent dust, contaminans and insects from getting inside the package.

## 3.13    High-Performance Fibres

High-performance fibre-reinforced cementitious composites are a group of fibre-reinforced cement-based composites which possess the unique ability to flex and self-strengthen before fracturing. High-performance fibres, used in fabric applications ranging from bulletproof vests to trampolines, must have a sufficient number of chemical and physical bonds for transferring the stress along the fibre. To limit their deformation, the fibres should possess high stiffness and strength. Stiffness is brought about by the degree to which the chemical bonds are aligned along the fibre axis. In fibre-reinforced composites, the fibres are the load-bearing element in the structure, and they must adhere well to the matrix material.

In a sense, all fibres except the cheapest commodity fibres are high-performance fibres. High-performance fibres offer special properties due to the demands of the respective application. These demands cover properties such as high tension, high elongation and high resistance to heat and fire and other environmental attacks. They are generally niche products, but some are produced in large quantities. The natural fibres like cotton, wool, silk, etc., have a high aesthetic appeal in fashion fabrics (clothing, upholstery, carpets). Until 100 years ago, they were also the fibres used in engineering applications – what are called technical or industrial textiles. With the introduction of manufactured fibres (rayon, acetate, nylon, polyester, etc.) in the first half of the 20th century, not only were new high-performance qualities available for fashion fabrics, but they also offered superior technical properties. For example, the reinforcement in automobile tyres moved from cotton cords in 1900, to a sequence of improved rayons from 1935 to 1955, and then to nylon, polyester and steel, which dominate the market now. A similar replacement of natural and regenerated fibres by synthetic fibres occurred in most technical textiles.

The combination of moderately high strength and moderately high extension gives a very high energy to break, or work of rupture. Good recovery properties mean that they can stand repeated high-energy shocks.

In this respect, nylon and polyester fibres are unchallenged as high performance fibres, though their increase in stiffness with rate of loading reduces their performance in ballistic applications. It is notable that polyester has proved to be the fibre of choice for high-performance ropes with typical break loads of 1500 tonnes, used to moor oil-rigs in depths of 1000–2000 m. The high-stretch characteristics of elastomeric fibres, such as Lycra, have an undeveloped potential for specialized technical applications. However, because of their large-scale use in general textiles, these fibres are dealt with in another book in this series.

## 3.13.1    High-Performance Fibres

1.  Glass fibre
2.  Carbon fibre (CF)
3.  Aramid fibre
4.  Polybenzimidazole fibre
5.  Polyphenylenebenzobisoxazole and polyimide fibres
6.  Polyphenylene sulphide fibre

7. Melamine fibre

8. Fluoropolymer (polytetrafluoroethylene)

9. HDPE

10. Ceramic fibres

11. Chemically resistant fibres

12. Thermally resistant fibres

## 3.13.2    Glass Fibre

Glass fibre is the oldest, and most familiar, high-performance fibre. Fibres have been manufactured from glass since the 1930s. Although early versions had high strength, they were relatively inflexible and not suitable for several textile applications. Today's glass fibres offer a much wider range of properties and can be found in many end-uses, such as insulation batting, fire-resistant fabrics and reinforcing materials for plastic composites. Items such as bathtub enclosures and boats, often referred to as 'fibreglass' are, in reality, plastics (often cross-linked polyesters) with glass fibre reinforcement. And, of course, continuous filaments of optical quality glass have revolutionized the communications industry in recent years (Fig. 3.21).

**Fig. 3.21**    Glass Fibre

### 3.13.2.1    Types of Glass Fibre

As to the raw material glass used to make glass fibres or non-wovens of glass fibres, the following classification is known:

1.  *A-glass*: With regard to its composition, it is close to window glass. In the Federal Republic of Germany, it is mainly used in the manufacture of process equipment.

2.  *C-glass*: This kind of glass shows better resistance to chemical impact.

3.  *E-glass*: This kind of glass combines the characteristics of C-glass with very good insulation to electricity.

4.  *AE-glass*: Alkali-resistant glass.

Generally, glass consists of quartz sand, soda, sodium sulphate, potash, feldspar and a number of refining and dying additives. The characteristics, with them the classification of the glass fibres to be made, are defined by the combination of raw materials and their proportions. The cross section of the textile glass fibres mostly shows a circular structure.

### 3.13.2.2    Properties of Glass Fibre

Glass fibres are useful because of their high ratio of surface area to weight. However, the increased surface area makes them much more susceptible to chemical attack. By trapping air within them, blocks of glass fibre make good thermal insulation, with a thermal conductivity of the order of 0.05 W/(mK).

The strength of glass is usually tested and reported for 'virgin' or pristine fibres those which have just been manufactured. The freshest, thinnest fibres are the strongest because the thinner fibres are more ductile. The more the surface is scratched, the less the resulting tenacity. Because glass has an amorphous structure, its properties are the same along the fibre and across the fibre. Humidity is an important factor in the tensile strength. Moisture is easily adsorbed and can worsen microscopic cracks and surface defects and lessen tenacity.

In contrast to CF, glass can undergo more elongation before it breaks. There is a correlation between bending diameter of the filament and the filament diameter. The viscosity of the molten glass is very important for manufacturing success. During drawing (pulling of the glass to reduce fibre circumference), the viscosity should be relatively low. If it is too high, the fibre will break during drawing. However, if it is too low, the glass will form droplets rather than drawing out into fibre.

### 3.13.2.3    Glass Fibre Manufacturing Processes

After the initial process of melting glass and passing it through spinnerets, continuous filaments or staple fibres of glass are manufactured by two different methods.

(a) *Continuous filament process*: In this process, continuous filaments of indefinite length are produced. The molten glass passes through spinnerets having hundreds of small openings. These strands of multiple filaments are carried to winder revolving at very high speed of more than 2 miles/km. This process draws out the fibres in parallel filaments of the diameter of the openings. A sizing or a binder is applied to facilitate the twisting and winding process and to prevent breakage during yarn formation. After winding, filaments are further twisted and plied to make yarns by methods similar to those for making other continuous filament yarns. The sizing is removed through volatizing in an oven. These yarns are used for making such items as curtains and drapes.

(b) *Staple fibre process*: Fibres with long-staple qualities are manufactured through staple fibre process. There are many methods for producing such fibres.

In one of such methods, the molten glass flows through the small holes of bushing, where jets of compressed air shake the thin streams of molten glass into fine fibres. These fibres vary in length ranging from 8 to 15 inches. The fibres fall through a spray of lubricant and a drying flame onto revolving drum where they form into a thin web. These fibres in the form of web are gathered from the drum into a sliver. Yarn is then made from this sliver by similar methods that are adopted for making cotton or wool yarns. These yarns are used for fabrics for industrial purposes where insulation is required.

In yet another method, the ends of the glass rods are melted from which drops of glass fall away drawing off glass filaments after them onto a speedily revolving cylinder where they are wound parallel to each other. A web of sliver is formed if the cylinder moves sideways. Sometimes, the staple may be thrown off the cylinder onto a stationary sieve, where it forms a sliver. In either conditions, the sliver is then converted into spun yarn. The staple fibre, if subjected to oven, is compressed to the desired thickness and the binder which was earlier applied, is cured. This permanently binds the fibres.

### 3.13.2.4    Production

The subsequent manufacture of glass fibres may be executed to the direct melting process. However, in most cases, glass rods or balls are made first which then may undergo a variety of further processes.

(a) *Nozzle drawing*: As can be seen in Fig. 3.22(a) and (b), the glass fed in is melted in a heated melt tub at 1250–1400°C. Then, it emerges at the bottom of the melt tub from nozzle holes of 1–25 mm diameter, and it is taken off and drawn. The filaments solidify and are finished and wound. One can find them in the shops as various kinds of 'glass silk'. To make them into webs, the filaments are cut to length (mostly, between 6 and 25 mm).

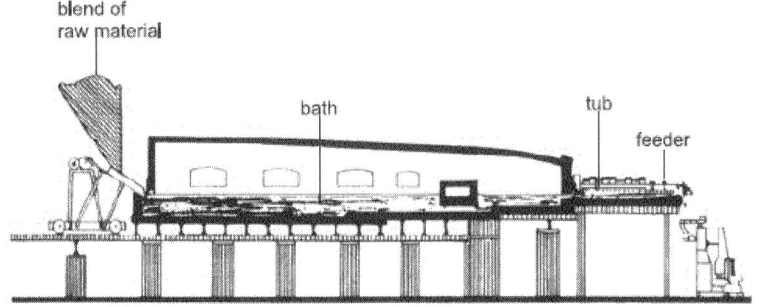

**Fig. 3.22(a)**    Manufacture of Glass Melt

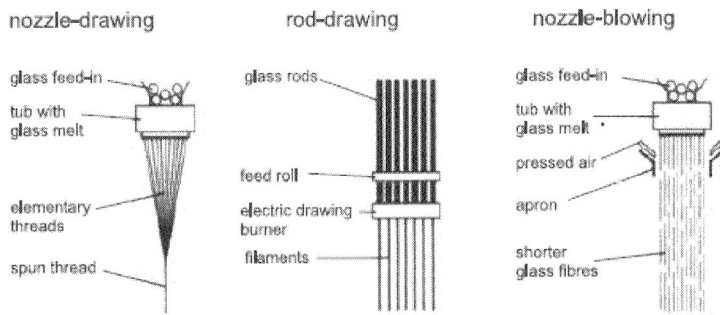

**Fig. 3.22(b)**    Glass Fibre Making Process

(b) *Nozzle blowing*: The same as with nozzle drawing, glass balls are melted in the tub. The melt emerging from the nozzle holes is then taken by pressed air, which draws the liquid glass so as to make fibres of 6–10 µm diameter. A fluttering effect is caused by the flow of pressed air, which results in fibres of lengths from 50 to 300 mm. A lubricant is put on and the fibres are laid down on a sieve drum which sucks them in. The dry web received is held together by the long fibres, the short ones lying in between them as a filling material. Then, the slivers of glass fibre material are cut.

(c) *Rod drawing*: By means of a burner, bundles of glass rods are melted at their bottom ends. This results in drops which, as they fall down, draw filaments after them. The filaments are taken by a rotating drum, a squeegee laying them down onto a perforated belt. Thus, a dry web is received which can be wound as glass fibre slivers. Machine performance being limited by the number of glass rods fed in, the rotating drum may be combined with nozzle drawing, which results in drum drawing. This multiplies machine performance. The dry web is again laid down onto a perforated belt and solidified or, after winding it so as to receive slivers, cut for further processing on machines producing wetlaid non-wovens. Using and processing glass fibres is not without any problems. For example, fine pieces of broken fibres may disturb if the work place is not well prepared for the purpose. Using the non-wovens to manufacture glass-fibre reinforced plastics, it is important the surface of the plastic material is fully even. Ends of fibre looking out may be pulled out or loosened by outward stress (temperature, gases, liquids), which may influence material characteristics. In some cases, it is advisable to cover up such layers of glass fibre with suitable chemical fibres.

### 3.13.2.5    Uses of Glass Fibre or Glass Yarn

Glass fibre is manufactured in a wide range of fine diameters. Some of them are so fine that they can be seen only through a microscope. This quality of fineness contributes greatly to the flexibility of glass fibres. Various manufacturers produce different types of glass fibres for different end-uses. Glass fibres then are used for various purpose.

1. For making home furnishings fabrics
2. For making apparels and garments
3. For the purpose tires and reinforced plastics

There are certain glass fibres that can resist heat up to 7200°C and can withstand forces having speed of 15,000 miles/h. These types of glass fibres are used as

1. filament windings around rocket cases,
2. nose cones,
3. exhaust nozzles and
4. heat shields for aeronautical equipment.

Some other types of glass fibres are embedded into various plastics for strength. These are used in

1. boat hulls and seats,
2. fishing rods and
3. wall panelling.

Some other types of glass fibres are used for reinforcing electrical insulation. Yet other types are used as batting for heat insulation in refrigerators and stoves.

## 3.13.3    Carbon Fibre

The CF, alternatively graphite fibre, carbon graphite, is a material consisting of fibres about 5–10 μm in diameter and composed mostly of carbon atoms. CF may also be engineered for strength. CF variants differ in flexibility, electrical conductivity, thermal and chemical resistance. Altering the production method allows CF to be made with the stiffness and high strength needed for reinforcement of plastic composites, or the softness and flexibility necessary for conversion into textile materials. The primary factors governing the physical properties are degree of carbonization (carbon content, usually greater than 92% by weight) and orientation of the layered carbon planes. Fibres are produced commercially with a wide range of crystalline and amorphous content (Fig. 3.23).

**Fig. 3.23**   Carbon Fibres

Because carbon cannot readily be shaped into fibre form, commercial CFs are made by extrusion of some precursor material into filaments, followed by a carbonization process to convert the filaments into carbon.

### 3.13.3.1   From *Polyacrylonitrile*

- Polyacrylonitrile (PAN) contains highly polar C–N groups that are randomly arranged on either side of the chain.
- Carbon filaments are wet spun from a solution of PAN and stretched at an elevated temperature during which the polymer chains are aligned in the filament direction. Then the filaments are heated at 200–300°C for a few hours.

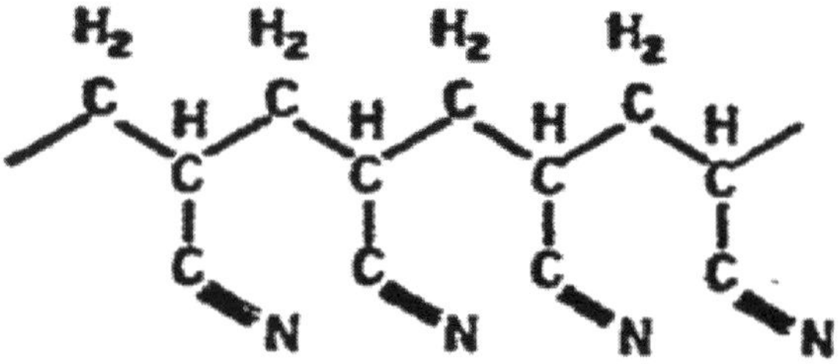

**Fig. 3.24**   Structure of Carbon Fibre

- At this stage the C–N groups located on the same side combine to form a more stable and rigid ladder like structure and some of the CH2 groups are oxidized (Fig. 3.24).
- In the next stage, the PAN filaments are carbonized by heating at a controlled rate between 1000 and 2000°C in an inert atmosphere.
- Tension is maintained on the filament to prevent shrinkage and to improve molecular orientation.
- Subsequently the carbonized filaments are heated above 2000°C, where their structures become more oriented and turns towards a true graphite form with increasing heat treatment temperature.
- At this stage, the graphitized filaments attain a high tensile modulus, but their tensile strength may be relatively low.
- Tensile strength can be increased by hot stretching above 2000°C.

### 3.13.3.2    From Pitch

Pitch is a by-product of petroleum refining and is a lower cost raw material than PAN.

- The carbon atoms in pitch are arranged in low molecular weight aromatic ring patterns.
- Heating to temperature above 300°C polymerizes these molecules into long two-dimensional sheet-like structure.
- The highly viscose state of pitch at this stage is called mesophase.
- Pitch filaments are produced by melt spinning the mesophase pitch through a spinneret.
- The filaments are cooled to freeze the molecular orientation and then heated between 200 and 300°C in oxygen atmosphere to stabilize them and make them infusible.
- In the next step, the filaments are carbonized at 2000°C.
- Rest of process of transferring the structure to graphitic form is similar to that followed for PAN precautions.

### 3.13.3.3    Uses

The CFs are used in the following fields: fishing rods, fishing nets, aerospace, sporting goods, automobiles, wind turbine blades, military, medical applications and many more fields.

## 3.13.4    Aramid Fibre

All fibres used in polymer engineering composites can be divided into two categories, namely synthetic fibres and natural fibres. Synthetic fibres are the most common. Although there are many types of synthetic fibres, glass, carbon and aramid fibres represent the most important. Kevlar is an aromatic polyamide or aramid fibre introduced in early 1970s by DuPont. It was the first organic fibre with sufficient tensile strength and modulus to be used in advanced composites. It has approximately five times the tensile strength of steel with a corresponding tensile modulus. Originally developed as a replacement for steel in radial tires, aramid is now used in a wide range of applications. It is a trade name of aramid fibre (Vigotsky, 2002).

### 3.13.4.1    History

First time aramid fibres commercially introduced by an American company DuPont in 1960 with trade name of Nomex. These Nomex fibres were well known due to their good thermal and electrical insulations properties. In 1971, DuPont introduced a much higher tenacity and modules fibre with trade name of Kevlar. Scientists in the fields of liquid crystals, polymers, rheology and fibre processing, as well as process and system engineers, spent several years prior and during the early stage of its market introduction establishing the basics and fundamental understanding necessary to take full advantage of this new class of high-performance materials.

### 3.13.4.2    Basic Structure and Chemical Composition of Aramid Fibres

The monomers of aramid fibres are consisting of 1,4-phenyl-diamine (para-phenylenediamine) and terephthaloyl chloride. The result is polymeric aromatic amide with altering benzene ring and amide groups. When they produced these polymer strand aligned randomly. Technically, aramid fibres are long-chain synthetic polyamides. Aramid fibres have extremely high tensile strength, which is why they are commonly used in armour and ballistic protection applications. With a distinctive yellow colour, aramid fibres are frequently used in advanced composite products which require high-strength and lightweight properties.

The chemical composition of aramid is poly para-phenylene diamine and terephthaloyl (PPD-T) and it is more properly known as a para-aramid. It is oriented para-substituted aromatic units. Aramids belong to the family of nylons. Common nylons, such as nylon 6,6, do not have very good structural properties, so the para-aramid distinction is important. Aramid fibres like Nomex or Kevlar, however, are ring compounds based on the structure of benzene as opposed to linear compounds used to make nylon. The aramid ring gives thermal aramid stability, while the para structure gives it high strength and modulus. Like nylons, aramid filaments are made by extruding the precursor through a spinneret. The rod form of the para-aramid molecules and the extrusion process make Kevlar fibres anisotropic – they are stronger and stiffer in the axial direction than in the transverse direction. In comparison, graphite fibres are also anisotropic, but glass fibres are isotropic (Fig. 3.25).

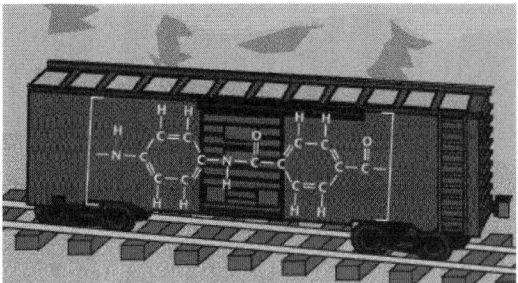

**Fig. 3.25**    Chemical Composition of Kevlar

It is made from a condensation reaction of PPD-T chloride. The result-ant aromatic polyamide contains aromatic and amide groups which makes them rigid rod-like polymers. The rigid rod-like structure results in a high glass transition temperature and poor solubility, which makes fabrication of these polymers, by conventional drawing techniques, difficult. Instead, they are melt spun from liquid crystalline polymer solutions as described later. The Kevlar fibre is an array of molecules oriented parallel to each other like a package of uncooked spaghetti. This orderly, untangled arrangement of molecules is described as a crystalline structure. Crystallinity is obtained by a manufacturing process known as spinning, which involves extruding the molten polymer solution through small holes.

When PPD-T solutions are extruded through a spinneret and drawn through an air gap during fibre manufacture, the liquid crystalline domains can orient and align in the flow direction. Kevlar can acquire a high degree of alignment of long, straight polymer chains parallel to the fibre axis. The structure exhibits anisotropic properties, with higher strength and modulus in the fibre longitudinal direction than in the axial direction. The extruded material also possesses a febrile structure. This structure results in poor shear and compression properties for aramid composites. Hydrogen bonds form be-tween the polar amide groups on adjacent chains, and they hold the individ-ual Kevlar polymer chains together (Fig. 3.26).

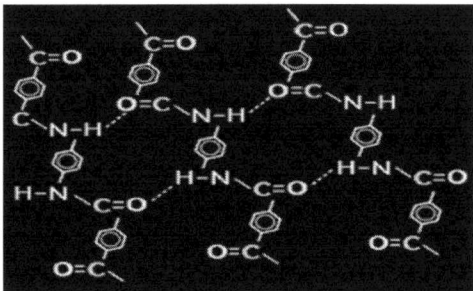

**Fig. 3.26**    Hydrogen Bonds Form Between the Polar Amide Groups

### 3.13.4.3    Types of Aramid Fibres

There are two main types of aramid fibres.

1. Meta-aramid
2. Para-aramid

The term meta and para refer to the location of chemical bonds in the structure of aramid fibres. The chemical bonds of a para-aramid fibres are more aligned in the long direction of the fibres. The meta-aramid fibres are not aligned. They are in zig-zag pattern, therefore, they are not developed the higher tensile strength of the para-aramid bonds.

(a) *Meta-aramid*:  Fibres made from the meta aramid have the excellent thermal, chemical and radiation resistance and are make the fire retardant textiles such as outer wear for fire fighters and racing car drivers. Nomex and Teijiconex are examples of meta aramids.

(b) *Para-aramids*: Fibres which are made from the para-aramid have higher strength. These are more commonly used in fibre reinforcement plastics for civil engineering structures, stress skin panels and other highly tensile strength applications. Kevlar and Technora are examples of para-aramid fibres (Hearle, 2001; Properties of Aramid fibres, 2015).

### 3.13.4.4    Different Trade Name of Aramid Fibres

Aramid fibres are available with different trade names. There properties are determine by the manufacturing process, conditions in which fibres are prepared and end-uses. Different trade names of aramid fibres are Kevlar, Technora, Tawron, Nomex, etc.

### 3.13.4.5    Manufacturing Process of Aramid Fibres

The polymer polymetaphenylene isophthalamide is used to make meta-aramids and the polymer $p$-phenylene terephthalamide to make para-aramids. Because the aramids decompose before they melt, they are produced by wet and dry spinning methods. Sulphuric acid is the normal solvent used in the spinning processes. In wet spinning, a strong solution of the polymer, which also contains inorganic salts, is spun through a spinneret into weak acid or water. In this bath, the salts leach out. In the dry spinning process, the salts are more difficult to remove and this process is only used to produce the weaker meta-aramid fibres. In both processes, post-treatment of the fibres by additional drawing is used to optimize fibre properties. Aramid products are available as filament yarn, staple fibre or pulp (Fig. 3.27).

## Aramid

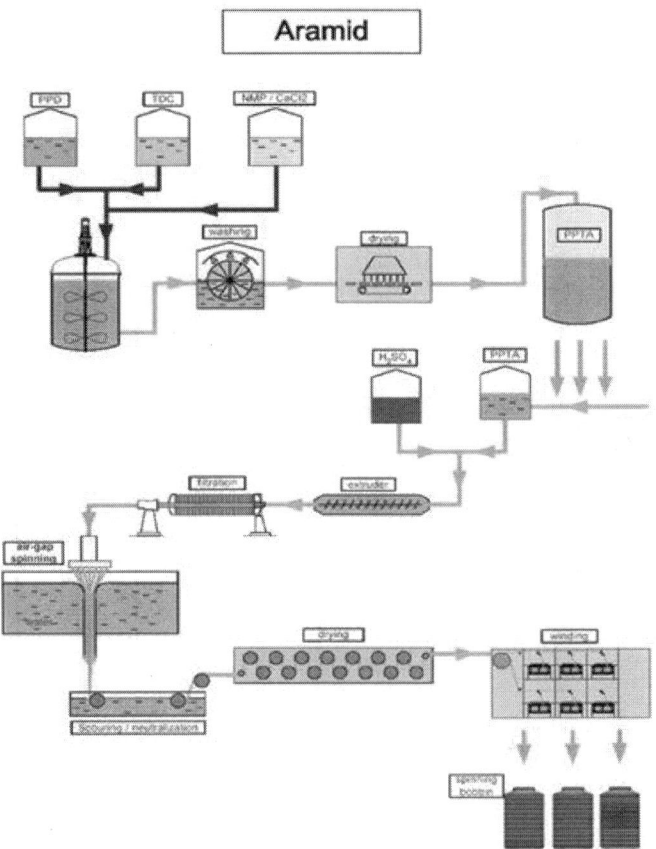

**Fig. 3.27**  Man-made Fibres (2015)

### 3.13.4.6  Characteristics of Aramid Fibres

1. *Fibre Structure*: In aramid fibres, a series of synthetic polymers in which repeating units have large phenyl rings are linked together by amide groups. Amide groups (CO–NH) form strong bonds that are resistant to solvents and heat. Phenyl rings are bulky six-sided groups of carbon and hydrogen atoms that prevent polymer chains from rotating and twisting around their chemical bonds.

2. *Fibre Properties*: Aramid fibres have medium to ultra-high strength, medium to low elongation and moderately high to ultra-high modulus with the densities of 1.38 g/cm$^3$. Heat-resistant and flame-retardant fibres contain high proportion or meta-oriented phenylene rings, whereas ultra-high-strength, high-modulus fibres contain mainly para-oriented phenylene rings.

3. *Chemical Properties*: All aramids contain amide links that are hydrophilic. However, not all aramid products absorb the same moisture. The PPD-T fibre has very good resistance to many organic solvents and salt, but strong acids can cause substantial loss of strength. Aramid fibres are difficult to dye due to their high $T_g$. In addition, the aromatic nature of para-aramid is responsible for oxidative reactions when exposed to UV light that leads to a change in colour and loss of some strength.

4. *Thermal Properties*: Aramid fibres do not melt in the conventional sense but decompose simultaneously. They burn only with difficulty because of limited oxygen index values. It should be mentioned that at 300°C, some aramid types can still retain about 50% of their strength. Aramid fibres show high crystallinity which results in negligible shrinkage at high temperature.

5. *Mechanical Properties*: Aramid yarn have breaking tenacity of 3045 MPa, in other words more than five times than this steel (under water, aramid is four times stronger) and twice than this of glass fibre or nylon. High strength is result of its aromatic and amide group and high crystallinity. Aramid retains strength and modulus at temperatures as high as 300°C. It behaves elastically under tension. When it comes to severe bending, it shows non-linear plastic deformation. With tension fatigue, no failure is observed even at impressively high loads and cycle times. Creep strain for aramid is only 0.3% (Fig. 3.28).

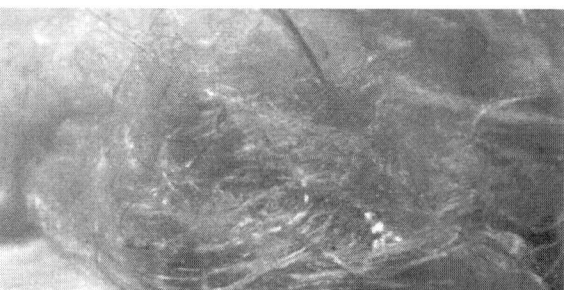

**Fig. 3.28**  Aramid Fibres

# References

1. Chandra Ray, S. Jute – the golden fibre of India, Modern Textile Journal, Jan–Feb, 2004, pp. 70–82.
2. Sur, D. Understanding Jute Yarn. Anindita Sur, Kolkata, India, 2005.

3. Jute and Allied Fibre Updates, Production and Technology. Central Research Institute for Jute and Allied Fibres (CRIJAF), Barrackpore, West Bengal, India, 2008.

4. Roul, C. The International Jute Commodity System. Northern Book Centre, New Delhi, India, 2009.

5. Sobha, M.A. Jute Agriculture, Jute Basics. International Jute Study Group, Dhaka, Bangladesh, 2010.

6. Sur, D., Nurul Amin, M. Physics and Chemistry of Jute, Jute Basics. International Jute Study Group, Dhaka, Bangladesh, 2010.

7. Abdullah, A.B.M. Jute Products, Jute Basics. International Jute Study Group, Dhaka, Bangladesh, 2010.

8. Aleem Ahmed, M. Textile Fiber, Applied Chemistry Research Centre.

9. California Environmental Protection Agency, Toxic Air Contaminant Identification List Summary for Fine Mineral Fibers. ARB/SSB/SES, Berkeley, CA, USA, 1997.

10. K domstift brandenburg, steglich, w leucht and w thumann, Use of natural fibres for facade elements and process for producing the same, EP 0700477, 13 March 1996.

11. www.gate.iisc.ernet.in/wp-content/.../Textile_Engineering_and_Fibre_Science.pdf

12. https://www.ncbi.nlm.nih.gov/pmc/articles/PMC1258937/pdf/biochemj01181-0031.pdf

13. https://www.scribd.com/document/246214411/Textile-fiber-Intro-pdf

14. 72.10.50.51/wp-content/uploads/2011/03/Polymers-In-Textile-Fibers.pdf

15. www.dressandtextilespecialists.org.uk/wp-content/uploads/2015/.../Fibres-Fabrics.pdf

16. www.troficolor.pt/en/download/troficolor-textile-processes-textile-fibres.pdf

17. www.fao.org/economic/futurefibres/fibres/jute/en/

18. https://www.intechopen.com/books/.../plant-fibres-for-textile-and-technical-applications

19. www.technicaltextile.gov.in/dnloads/Agrotech%20-%20Sector%20Presentation.pptx

20. http://news.bio-based.eu/polylactic-acid-pla-market-in-packaging-textile-agriculture

# 4

# Properties of Agro Textiles

In their simplest form, textiles have been used in agriculture for thousands of years to protect plants – as well as animals – against extreme climatic conditions. For instance, they offer shade, help to maintain soil humidity and increase soil temperature, and can also protect crops from insects and weeds.

For most end-uses, agro textiles require suitable tensile strength and good permeability characteristics with no significant deterioration under the influence of climatic extremes. The properties of agro textiles are dependent on the fibres from which they are produced and on the type and conditions of their manufacture.

However, advanced technical agro textile products have become commonplace only in the past two or three decades, with fibrous materials, particularly synthetics, beginning to replace plastics in certain applications. In particular, the use of non-wovens, especially spun bond fabrics, is increasing.

Man-made fibres are preferred for agricultural products than the natural fibres, mainly due to their favourable price performance ratio, ease of transport, space saving storage and long service life as well as properties.

## 4.1 General Properties Agro Textiles Products

### 4.1.1 Strength

The strength indicates the resistance sustained by fibres, yarns or fabrics to break when force is applied to them. The strength may be a tensile strength, bending strength, bursting strength, etc., as per the direction of application of force. Fibre strength and elongation have a direct relationship with yarn strength and elongation another thing being constant. A stronger fibre results into stronger yarn and same is true about fibre elongation.

The strength of any material is derived from the load it supports at the break and is thus a measure of its limiting load bearing capacity. Normally strength of a textile fibre is measured in tension when the fibre is loaded along its long axis and is designated as 'tensile strength'. The tensile strength of the textile fibre is measured as the maximum tensile stress in force per unit cross-sectional area or per unit linear density, at the time of rupture — called 'tenacity', expressed in terms of grams per denier or grams per tex units.

## 4.1.2 Uniformity

Uniformity means the evenness of the individual fibres in its length and diameter. A fibre possessing this property can produce reasonably even yarns. This is also important in connection with the strength of the resulting yarn. Uniform textile fibres should possess uniformity in their thickness and length. Unfortunately, none of the principal natural fibres like cotton or wool has the same length and diameter of the fibre in the same lot. Fibres in any specified qualities, grades or lots vary considerably in length and diameter. On the contrary, man-made staple fibres are more uniform as they are cut to the exact length after being spun and drawn, and even the diameter can be controlled within close tolerance limits during its manufacture.

## 4.1.3 Spinnability

Spinnability includes several physical properties each having an effect on the ability of the fibres to be spun into yarn. Staple fibres must have to be capable of taking a twist or being twisted (flexibility). They must have a certain degree of friction (cohesion) against one another to stay in place when the pull is applied to the yarn. They must also be able to take on whole special finishes for lubrication during spinning or to provide additional surface resistance to abrasion.

## 4.1.4 Flexibility

The fibre should be sufficiently pliable; then only it can wrap around another fibre during spinning. If fibre is stiffer, then it is less adaptable for textile use, for example, glass and metallic fibres.

## 4.1.5 Cohesiveness

Cohesiveness is the property of clinging or sticking together in a mass. Usually, rigid fibres have lower cohesiveness. It is generally assumed that a high degree of frictional resistance plays a part in the cohesiveness. It is the property of an individual fibre by virtue of which the fibres hold on to one another when the fibres are spun into yarn. This action is usually brought about by the high degree of frictional resistance offered by the surface of the fibres to separate one from the other. The wool fibres, for example, have a saw-toothed surface, so that the projecting edges on its surface, called scales, easily catch on to one another when several such fibres are twisted together during spinning. On account of this, fibres offer resistance when an attempt is made to pull them apart. Cotton fibres also possess irregular or rough surface. Further, due to the natural twist in the cotton fibre known as convolution, the fibres interlock themselves by friction when they are spun into yarns. The introduction of a crimp in synthetic fibre increases cohesion.

## 4.1.6 Resilience

Resilience is the springing back or recovery of a fibre when it is released from a deformation. The resistance to compression, flexing or torsion varies from fibre to fibre. Some fibres have a natural tendency to return to their original condition when any of the above-mentioned forces is applied, an important property where, for instance, recovery from creasing is required. Wool is outstanding in this respect by virtue of its natural characteristics, but cellulosic fibres may be modified in such a manner so as to greatly improve these properties. This springiness of a fibre or its mass resilience is highly desirable for carpet wool. Because of resiliency fibre/yarn/fabrics hold their shape, drape gracefully and do not wrinkle.

## 4.1.7 Capillarity and Porosity

Porosity can be defined as the ratio of the volume of air contained within the boundaries of the material to the total volume of a solid plus air or void, expressed as a percentage. Porosity facilitates the absorption of moisture, liquid lubricants, dyes, oils and steam by the fibres so as to thoroughly permeate the fibre. Porosity in a fibre is important in wet processing. The natural and man-made fibres differ greatly in respect of porosity which in turn affects other properties of fibres and consequently the processing of fibres during textile manufacture. In general, natural fibres have higher porosity than synthetic fibres.

## 4.1.8 Durability

Agro textile fibre should withstand processing treatments and should not be easily susceptible to physical, chemical and bacteriological attack, which may result in damage and decomposition. The durability of fabric to average wear and tear depends somewhat more on the elasticity, flexibility and resistance of the fibre and fabric, rather than the absolute strength of either fibre or fabric. If a fabric possesses these three properties, its products will absorb or counter more readily stresses and strains during strain. It will allow itself to be deformed with less resistance, thus reducing the chance of intermediate tearing or twisting. A raised fabric surface increases fabric resilience and provides longer resistance to abrasive surfaces, for example, mats, nets, fabrics, etc.

## 4.2    Specific Properties

### 4.2.1    Withstands Solar Radiation

Agro textiles are laid over the cultivated areas immediately after sowing or planting. For such application, agro textile has to withstand solar radiation with varying surrounding temperature. The warp-knitted nets are used to protect fields and greenhouses from the intense solar radiation for healthy plant growth and good harvest. The 'greenhouse effect' of the atmosphere is named by analogy to greenhouses which become warmer in sunlight. The explanation given in most sources for the warmer temperature in an actual greenhouse is that incident solar radiation in the visible, long-wavelength ultraviolet (UV) and short-wavelength infrared (IR) range of the spectrum passes through the glass roof and walls and is absorbed by the floor, earth and contents, which become warmer and re-emit the energy as longer-wavelength IR radiation (IRR). Glass and other textile materials used for greenhouse walls do not transmit IRR, so the IR cannot escape via radiative transfer. As the structure is not open to the atmosphere, heat also cannot escape via convection, so the temperature inside the greenhouse rises.

A greenhouse works primarily by allowing sunlight to warm surfaces inside the structure, but then preventing absorbed heat from leaving the structure through convection. The 'greenhouse effect' heats earth because greenhouse gases absorb outgoing radiative energy, heating the atmosphere which then emits radiative energy with some of it going back towards earth. Shading the roof of a greenhouse is usually performed by various conventional methods such as whitening the roof, external shade cloths, deploying plastic nets of various colours and movable refractive screens or curtains. However, it reduced the average greenhouse transmittance to solar radiation from 0.62 to 0.31.

The external shade cloth is usually applied by deploying wet or dry shade cloths on the outer surface of the greenhouse roof. An external or internal shade can also be obtained by using movable plastic nets, curtains or refractive screens applied above or below the roof of the greenhouse. All shading methods are to regulate the amount of solar energy entering the greenhouse and reduce the heating load in summer. Besides protecting plants against excessive heat load, shading significantly reduces the water requirement in arid regions. Disadvantage of shading system that used curtain or screen below the roof of the greenhouse is that when the curtain or screen is fully deployed, it will decrease the effectiveness of the natural roof ventilation and negatively affect the greenhouse microclimate. Moreover, presence of shading materials deployed in the greenhouse absorbs a portion of solar radiation, reemits it again in the greenhouse and reflects back a portion also inside the greenhouse. Therefore, the effect of internal shading on reducing the greenhouse air temperature is expected to be small. All the aforementioned shading methods significantly reduce solar radiation across the whole solar spectrum including the photosynthetically active radiations (400–700 nm) which is essential for plant growth.

## 4.2.2    Withstands Ultraviolet Radiation

Polyethylene (PE) is resistant to radiation in the visible range but UV radiation (UVR) leads to degradation of molecular chains. Hence when used as an outdoor material, PE is treated with the appropriate UV stabilizers. These are special types of carbon black which convert the UVR into thermal radiation. Good potential to reduce the impact of UVR on plants by light-absorbing or light-reflecting non-wovens (light permeability: 80–90% to allow photosynthesis to take place).

The trapping of the sun's warmth in a planet's lower atmosphere, due to the greater transparency of the atmosphere to visible radiation from the sun than to IRR emitted from the planet's surface. The UVR is a type of solar radiation with wavelengths between 100 and 400 nm. Culturing plants in greenhouses might have two adverse consequences: less radiation at visible wavelengths for photosynthesis and less or no UVR (due to glass or plastic covers that block UVR). This latter reduces the production of UV-induced phenolic substances.

Earth is constantly bombarded with enormous amounts of radiation, primarily from the sun. This solar radiation strikes the earth's atmosphere in the form of visible light, plus UV, IR and other types of radiation that are invisible to the human eye. The UVR has a shorter wavelength and a higher energy level than visible light, while IRR has a longer wavelength and a weaker energy level. For greenhouses in hot and sunny regions, scientists and companies have worked for many years to develop greenhouse covering systems able to reduce the heat load as well as the air temperature in the greenhouses.

## 4.2.3    Biodegradability

Natural fibres like wool, jute and cotton are also used where the biodegradability of product is essential. Natural polymer gives the advantage of biodegradation but has low service life when compared to the synthetics.

Agriculture and tourism are important sectors for a country's economy. Waste production is a problem for both sectors. In the agriculture sector, waste plastic materials have a negative impact on the environment and increase the quantity of waste to be disposed. In the tourism sector, the quantity of waste disposed in landfills has risen significantly in the last years, and planning in control and prevention is lacking.

The materials used for 'traditional' yarns can be both plastic, such as polypropylene (the most used plastic material for this application) and PE, and natural fibres, such as jute and raffia. Biodegradable and compostable alternative are mainly produced in polylactic acid.

## 4.2.4    High Potential to Retain Water

This is achieved by means of fibre materials which allow taking in much water and by filling in super absorbers. While non-wovens meant for the covering of plants show a mass per unit area of 15–60 g/m², values between 100 and 500 g/m² are reached with materials for use on embankments and slopes.

## 4.2.5    Protection Property

This property includes protection from wind and the creation of a microclimate between the ground and the non-wovens, which results in temperature and humidity being balanced out. At the same time, temperature in the root area rises, which causes earlier harvests. It also includes sufficient stiffness, flexibility, evenness, elasticity, biodegradability, dimensional stability and resistance to wetness. Fungicidal finish (up to 2% of the total mass), which avoids soil contamination.

## 4.2.6    Resistance to Microorganisms

Biodeterioration has been defined as 'any undesirable change in the properties of a material caused by the vital activities of organisms'. Under suitable conditions, microorganisms which inhabit soil, water and air can develop and proliferate on textile materials. These organisms include species of microfungi, bacteria, actinomycetes (filamentous bacteria) and algae. Textiles made from natural fibres are generally more susceptible to biodeterioration than the synthetic man-made fibres although textiles, in general, provide a very suitable living environment for many microorganisms.

Negative effect on the vitality of the microorganisms is generally referred to as antimicrobial. The degree of activity is differentiated by the term 'cidal' which indicates significant destruction of microbes and the term 'static' represents inhibition of microbial growth without much destruction. The activity which affects the bacteria is known as antibacterial and that of fungi is antimycotic. The antimicrobial substances function in different ways. In the conventional leaching type of finish, the species diffuse and poison the microbes to kill. This type of finish shows poor durability and may cause health problems. The non-leaching type or biostatic finish shows good durability and may not provoke any health problems. A large number of textiles with antimicrobial finish function by diffusion type. The rate of diffusion has a direct effect on the effectiveness of the finish. For example, in the ion exchange process, the release of the active substances is at a slower rate compared to direct diffusion and hence has a weaker effect. Similarly, in the case of antimicrobial modifications where the active substances are not released from the fibre surface and hence rendering them less effective. They are active only when they come in contact with microorganisms. These so called new technologies have been developed by considering the medical, toxicological and ecological principles.

The agro textile products are used in moist areas with ample opportunities for microorganisms to grow on them. To avoid any damage to the material as well as the plants, it is very important for such products to be microorganism resistant.

Antimicrobial resistance threatens the effective prevention and treatment of an ever-increasing range of infections caused by bacteria, parasites, viruses and fungi. Antimicrobial resistance is an increasingly serious threat to global public health that requires action across all government sectors and society.

## 4.2.7    Dimensional Stability

It is the ability of a material to maintain its essential or original dimensions while being used for its intended purpose. The degree to which a material maintains its original dimensions when subjected to changes in temperature and humidity. The dimensional stability of a fabric is a measure of the extent to which it keeps its original dimensions subsequent to its manufacture.

Agro textiles fabrics are found to be with high tensile strength and good dimensional stability which is one of the basic requirements, for example, protecting the plantations from extreme natural conditions. The results are products of improved quality, increased yields, fewer losses and decreased damages. This permits substantially reduced usage of weed killers and pesticides. This property prevents the fabric to loosen up while it is being used, as the loosening or change in dimensions of the material may lead to non-usability of the material.

## 4.2.8    Flexibility

In agro textiles, fabric structures are forms of constructed fibres that provide end-users a variety of free-form functional designs. Custom-made fabric structures are engineered and fabricated to meet worldwide structural, microbial-resistant, weather-resistant and natural force requirements. Fabric structures are considered as a sub-category of tensile structure to allow the usage of agro textiles in varying areas and places.

## 4.2.9    Tensile Strength

The tensile strength of a material is the maximum amount of tensile stress that it can take before failure, for example, breaking. There are three typical definitions of tensile strength: Yield strength - The stress a material can withstand without permanent deformation. Tensile testing of textiles provides the strength and elongation properties for both natural and man-made materials, such as cotton, carbon, polyester, nylon, glass and graphite. Textiles can be tensile tested in many forms, including single strands, yarns, webbing, woven, braided material and non-woven fabrics. The majority of textile fabric tensile testing is performed as either a grab test, in an effort to eliminate edge effects, or a strip test, including edge effects. The tensile strength of shade nets can be a deciding factor of its long-term durability and service life. Hence good tensile strength is necessary parameter for shade nets.

## 4.2.10    Abrasion Resistance

Abrasion resistance is the ability of a fabric to resist surface wear caused by flat rubbing contact with another material. The abrasion to which a shade net is subjected may be of the material itself (material to material) or stray animals. Abrasion of the shade net would result in holes through which animals and pests could enter the structure and harm the crops. Good abrasion resistance is required of shade nets.

## 4.2.11    Stable Construction

Woven fabrics are made up of a weft – the yarn going across the width of the fabric and a warp – the yarn going across the length of the loom. The yarns are interlocked together. The side of the fabric where the wefts are double backed to form a non-fraying edge is called the selvedge. The construction must be such that it must be stable for any application.

## 4.2.12    Tear Strength

Tear strength is the resistance of the fabric against tearing or force required to propagate the tear once it is initiated. The tear strength is required in high-performance applications as well as in the conventional textiles, that is, in the industrial applications, bullet proof jackets, tents, worker jeans, sacks, aesthetic apparel and many more applications. This is also important in the industrial textiles where heavy duty work is performed. High tear strength of textiles makes sure that the punctures in the fabrics do not propagates easily.

## References

1. Subramaniam, V., Poongodi, G.R., Veena Sindhuja, V. Agro-textiles: production, properties & potential, The Indian Textile Journal, Vol. 119, Issue 7, 2009, pp. 73–77.

2. Collier, B.J., Epps, H.H. Textile Testing and Analysis. Merrill Prentice Hall, New Jersey, 1999, ISBN 0-13-488214-8.

3. Booth, J.E. Principles of Textile Testing, 3rd ed. CBS Publishers & Distributors Pvt. Ltd., New Delhi, 1996, ISBN 81-239-0515-7.

4. Raul, J. Textile Testing. APH Publishing Corporation, New Delhi, 2005, ISBN 81-7648-748-1.

5. Carr, H., Pomeroy, J. Fashion Design and Product Development. Blackwell Science Ltd., London, 1996, ISBN 0-632-02893-9.

6. Angappan, P., Gopalakrishnan, R. Physical Testing. SSM Institute of Textile Technology, Komarapalayam, 1991.

7. http://www.technicaltextilesinfrance.com

8. http://www.textileroadmap.com

9. http://www.technicaltextile.net

10. http://textilelearner.blogspot.com

# 5

# Classification Based on Areas of Application

Technical textiles are reported to be the fastest growing sector of the textile industrial sector and account for almost 19% (10 million tonnes) of the total world fibre consumption for all textile uses. Agro textile is one of the growing areas of technical textiles. The volume of special textiles that are manufactured for agricultural applications is small compared to other areas of technical textiles. This does not mean that the use of textiles in agriculture is not significant.

Agro textile is an application of textile materials in the agriculture field. With the continuous increase in population worldwide, stress on agricultural crops has increased. Hence, it is necessary to increase the yield and quality of agro products. But it is not possible to meet fully with the traditionally adopted ways of using pesticides and herbicides. Today, agriculture and horticulture has realized the need of tomorrow and opting for various technologies to get higher overall yield, quality and tasty agro products.

According to the literature, 'textiles for agro textiles' can be classified based on three categories.

- Based on the area of application
- Based on fabric production technique
- Based on the products

Agro textiles can be classified according to areas of applications. These areas are broadly identified as:

- Agro textiles for crop production
- Agro textiles for horticulture, floriculture and forestry
- Agro textiles for animal husbandry and aquaculture
- Agro textiles for agro-engineering-related applications

## 5.1    Agro Textiles For Crop Production

The selection of agro textile product is depends on crop needs. Selection of the agro textiles is also greatly influenced by the geographical location. Some of the applications of agro textiles are discussed in the following sections.

## 5.1.1    Sunscreen Net

These are used to protect fields and greenhouses from the intense solar radiation for healthy plant growth and good harvest. Sunscreen nets with open mesh construction are used to control sunshine and amount of shade required. These net fabrics allow the air to flow freely. Hence, the excess heat does not built up under the screen (Fig. 5.1). Every plant has its own individual requirements for sunlight and shade under which it flourishes at its best. If shading is required, pale-coloured materials should be used as these uniformly reflect solar radiation. A range of products exist that offer shading from 30% up to almost total black out. White wash paints are another option that can be applied to reduce the amount of radiation entering the greenhouse. By providing the right balance with the correct grade of shade cloth, the optimum climatic conditions are created under which the plant's productivity is maximized. Under these optimum conditions, photosynthesis is enhanced and extremes of air and soil temperature are reduced and moderated. Even more importantly, plant leaf temperatures are lowered to the same level as surrounding air temperature and it is this mechanism which accounts for the improved productivity of the plant.

The shade cloth impacts on the level and quality of light available to the crop. Diffused light is better than direct light. Fluorescent and pigmented fabric can increase the proportion of good red light. Dust attracted to the cloth will reduce the transmission of radiation. Water droplets on the inside coverings have been shown to reduce light transmission by 8% and also block thermal radiation.

The following are the important benefits of a shade cloth:

- Reduces the amount of solar energy entering the greenhouse, which lowers plant stress.
- Reduces light intensity.
- Air movement is restricted, thus reducing wind damage to the crop and evaporation of soil moisture.
- Air beneath the shade cloth stays humid, which is further benefit to the plant.
- Shade cloth provides a physical barrier against hail and heavy rain from damaging the plants.
- The requirement of percentage shade cloth for different types of plant depends upon the geographical location and the prevailing climatic conditions in that region. For example, 30% and 40% shade cloths provide ideal conditions for germination of seeds and development of seedlings at this highly vulnerable stage of a plant's life cycle. 50% and 55% shade cloths are for flower cultivation and are recommended for general nursery stock.

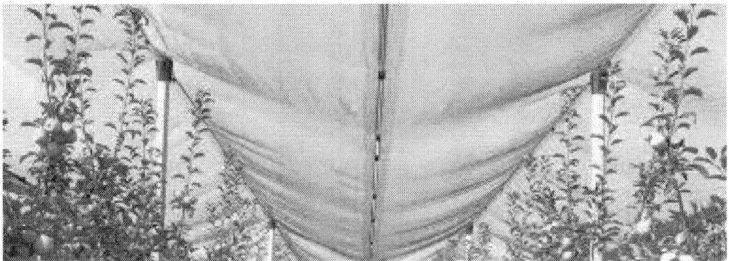

Fig. 5.1 Sunscreen Net

## 5.1.2 Bird Protection Nets

Knitted monofilament nets (open knitted nets for crop protection) offer effective passive protection of seeds, crops and fruits against damage caused by birds and a variety of pests. Open-mesh net fabrics are used as a means of protecting fruit plantation. The special open structure repels birds, provides minimal shading and excellent air circulation – allowing plants to flourish, while avoiding the risk of dangerous mould developing on the fruit (Fig. 5.2).

Fig. 5.2 Bird Protection Nets

## 5.1.3 Plant Net

Fruits, which grow close to the ground, can be kept away from the damp soil by allowing them to grow through vertical or tiered nets to keep the amount of decayed fruit to a minimum (Fig. 5.3). Nets are used for the protection of plants from excess sunlight, birds, insects, snow, hails wind, heavy rainfall, etc. The selection of nets varies as per the application with different mesh sizes or strength required to withstand weather conditions.

- Helps in cultivation of flower plants, foliage plants, medicinal and aromatic plants, vegetables and spices.
- Used for growing vegetables.
- Enhances yield in extreme climatic conditions.
- Helps in quality drying of various agro products.
- Used for pest/insect protection.

**Fig. 5.3**   Plant Net

## 5.1.4    Ground Cover

Ground cover is an extremely versatile landscaping and horticultural fabric for long-term weed control, moisture conservation and separation. It is mainly used in planted areas. It effectively suppresses competitive weed growth, conserves ground moisture, maintains a clean surface, protects from ultraviolet (UV) rays and creates a favourable environment for healthy plant growth. Ground covers can reduce the costs and minimizes undesirable herbicide use. Using this ground cover in display areas, nurseries and greenhouses will provide a clean, free draining and hard wearing surface. Fabric is regularly used to maintain a clean crop and to reduce maintenance and disease problems (Fig. 5.4).

**Fig. 5.4**  Ground Cover

## 5.1.5    Windshield

Windshields are used in farming to protect fruit plantations from wind and to prevent damage to plants. It also prevents plants being cooled by wind (Fig. 5.5).

**Fig. 5.5**  Windshields

## 5.1.6    Root Ball Net

It is extremely important for safe and speedy growing of young plants such that root system is not damaged when they are dug up, transported or replanted. Normally the root balls are wrapped in cloth. Elastic net tubes are alternative to this. When the plants are transplanted, the nets on the outside do not have to be removed since the roots can protrude through the nets (Fig. 5.6).

**Fig. 5.6**   Root Ball Net

### 5.1.7    Insect Meshes

Clearly, woven and knitted polyethylene monofilament meshes exclude harmful insects from greenhouses and tunnels or keep pollinating insects inside. The fine woven screens protect plants from insect attack (without the use of insecticides) (Fig. 5.7).

**Fig. 5.7**   Insect Meshes

### 5.1.8    Mulch Mat

Mulch mats are used to suppress weed growth in horticulture applications. It covers the soil, blocking of light and preventing the competitive weed growth around seedlings. It also reduces the need for herbicides required for weed control. Needle punched non-woven and black plastic sheet are used for this application. Biodegradable and non⊐biodegradable types of mulch mats are available (Fig. 5.8).

**Fig. 5.8**  Mulch Mat

## 5.1.9    Monofil Nets

Tough, knitted monofil nets are used as windbreak fences and shading/privacy screens. A suitable windbreak, set at a right angle to the prevailing wind, will protect plants against the harmful effects of blustery weather – which can break young branches, damage flowers and cause leaves to dry or tear. The nets also protect against frosts and help enhance the microclimate. This not only safeguards the current harvest but also benefits future crops, since the woody part of the plant are protected too (Fig. 5.9).

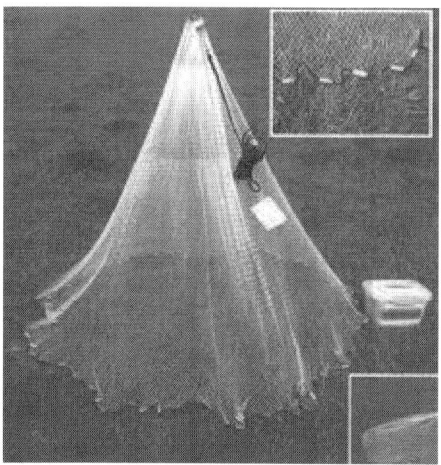

**Fig. 5.9**  Monofil Nets

## 5.1.10 Cold and Frost Control Fabrics

Cold and frost fabric can be laid directly on the plants, unlike plastic covers that can attract frost, and burn any leaf that touches them. These fabrics protect the plant from frost kill during unexpected late cold snaps and unexpected early ones (Fig. 5.10).

**Fig. 5.10** Cold and Frost Control Fabrics

## 5.1.11 Nets for Covering Pallets

For safe transportation of fruits and vegetables to the market, the boxes are covered with large mesh nets and pallets to stop the boxes being turned upside down. This prevents damage of goods during transportation (Fig. 5.11).

**Fig. 5.11** Nets for Covering Pallets

## 5.1.12    Packing Materials for Agricultural Products

Nets can be used for packaging of farm products for many end-uses. It includes packing sacks for vegetables, tubular packing nets for fruits and wrappers for Christmas trees. Net structures are preferred because of their high strength, low weight, air permeability and cheapness (Fig. 5.12).

**Fig. 5.12**    Packing Materials for Agricultural Products

## 5.2    Agro Textiles for Horticulture, Floriculture and Forestry

Application of textile materials in horticulture is growing fast. Nets, non-woven mats, movable screens for glass/poly houses, non-woven sheets, mixed bed for mushrooms, cordage and strings are used in horticulture. Nets are also used for protection against hailstorms, intense sunrays, etc. Non-woven sheets are used in the field to protect young plants such as strawberries, potatoes and lettuce from extreme cold weather, night frost and viruses. Light-resistant woven and non-woven polyester fabrics are used in the inside of greenhouse to protect the plants from extreme hot or cold conditions. They are also used on the outside of the greenhouses as screen to control sunlight (Figs. 5.13 and 5.14).

## 5.2.1      Horticulture

Horticulture is the branch of agriculture that deals with the art, science, technology and business of growing plants. It includes the cultivation of medicinal plants, fruits, vegetables, nuts, seeds, herbs, sprouts, mushrooms, algae, flowers, seaweeds and non-food crops such as grass and ornamental trees and plants. It also includes plant conservation, landscape restoration, landscape and garden designing, construction and maintenance, and arboriculture. Inside agriculture, horticulture contrasts with extensive field farming as well as animal husbandry.

Horticulturists apply their knowledge, skills and technologies to grow intensively produced plants for human food and non-food uses and for personal or social needs. Their work involves plant propagation and cultivation with the aim of improving plant growth, yields, quality, nutritional value and resistance to insects, diseases and environmental stresses. They work as gardeners, growers, therapists, designers and technical advisors in the food and non-food sectors of horticulture. Horticulture even refers to the growing of plants in a field or garden. Textiles play a vital role in the field of horticulture. Its significant role is development of various fabrics for different purposes.

Application of textile materials in horticulture is growing fast. Nets, non-woven mats, movable screens for glass/poly houses, non-woven sheets, mixed bed for mushrooms, cordage and strings are used in horticulture. Trees are covered by nets so as to protect the expensive fruits from birds. Nets are also used for protection against hailstorms, intense sunrays, etc. Light-resistant woven and non-woven polyester fabrics are used in the inside of greenhouses to protect the plants from extreme hot or cold conditions. They are also used on the outside of the greenhouses as screen to control sunlight. Fabric protective greenhouses provide virus-free cultivation of young plants. Non-woven sheets are used in the field to protect young plants such as strawberries, potatoes and lettuce from extreme cold weather, night frost and viruses. Nylon fabric beds are used to grow mushrooms. A mixture of horse manure and compost is laid on the fabric, which is wrapped around rollers to easily remove the mixture.

**Fig. 5.13**   Horticulture Net

## 5.2.2    Floriculture

Floriculture, or flower farming, is a discipline of horticulture concerned with the cultivation of flowering and ornamental plants for gardens and for floristry, comprising the floral industry. The development, via plant breeding, of new varieties is a major occupation of floriculturists.

Floriculture crops include bedding plants, houseplants, flowering garden and pot plants, cut-cultivated greens and cut flowers. As distinguished from nursery crops, floriculture crops are generally herbaceous. Bedding and garden plants consist of young flowering plants (annuals and perennials) and vegetable plants. They are grown in cell packs (in flats or trays), in pots or in hanging baskets, usually inside a controlled environment and sold largely for gardens and landscaping.

**Fig. 5.14**   Floriculture Net

Some of the agro textiles that are used frequently for horticultural and floriculture use are as follows:

- Hail protection fabrics
- Mulch net
- Rain protection fabrics
- Wind control fabrics
- Harvesting nets

### 5.2.3    Forestry

Woven technical wire mesh type is used in various styles for fencing the cultivated areas under forest control. The technical wire meshes can separate bulk grain sizes of gravel, grit, cinder, sand and many more. The wire meshes are made of high tensile strength wires with a zinc coating. This wire mesh type is perfect for forest all areas mainly for ornamental birds, poultry, birds of prey or other types of avifauna. The farming wire mesh comes with pre-finished sides for easier weaving and integration into larger sections.

The nettings are characterized by a high tensile strength and are impact resistant. Fencings made of those nettings are easy to assembly and enable quick and efficient fencing of premises. They are applied for fencing woodland areas, motorways, railway tractions, also held applicable in agriculture, for fencing pastures and runs for farm animals as well as for protecting arable crops (Fig. 5.15).

**Fig. 5.15**   Fences Used for Forestry

## 5.3 Agro Textile for Animal Husbandry and Aquaculture

Nylon and polyester identification belts are used for cows. Textile nets are used to support the large udders. Non-woven fabrics are used to filter the milk in automatic milking systems and as an underlay to reduce mud on cattle paths and trails. Agro textile is not only used for sheltering the animal from the low temperature at some stage in the wintertime but also during the hot summer months. It assists safeguard animals from high infrared (IR) rays and hot. In poultry house freshening will be done by two sets of changeable side windows, covered with mesh fabric.

Agro technical textiles are used in farming, animal husbandry and horticulture to control the hazardous influences of environmental and climactic factors on crop production and cattle breeding, regulate nutrient level intake of plants and assist in process and post-harvest operations (Fig. 5.16a).

Fishnets are key technical textiles used in fishing industry. Fishing nets are knitted fabrics used for oceanic and domestic fishing by fisherman, fishing trawlers and boats. The characteristics and specifications of fishnets vary based on the method adopted for fishing. Agro textiles, in the form of nets, ropes and lines, have also been used extensively in the fishing industry. Fishnets are used for fishing and in fish farming. Warp knitted knotless nets result in low energy expenditure when the net is used for fishing. They are mainly produced from nylon monofilament, multifilament or high-density polyethylene (HDPE) (Fig. 5.16b).

Fishnets (fish knitted fabrics) are used for catching fish by fishermen, in fishing boats and fishing trawlers. The mesh size ranges from 10 to 250 mm for usage in different fishing gears, which in turn depends on the fishing craft used. The yarn used for fishnet manufacture is nylon monofilament, nylon multifilament or HDPE. Fishnets are being manufactured by using power-driven net-making machines in small- and medium-scale sectors and by hand knitting in the cottage sector.

The major end-user sectors for fishing nets are Fisheries Development Corporations, Fishermen's Cooperatives, Private Sector trawling companies, etc. The uneconomic size of the fishnet manufacturing units and also limitations on range and type of the production capability do not enable them to sell to the world market. The slow growth of the domestic fishing industry limits the size of their market at home.

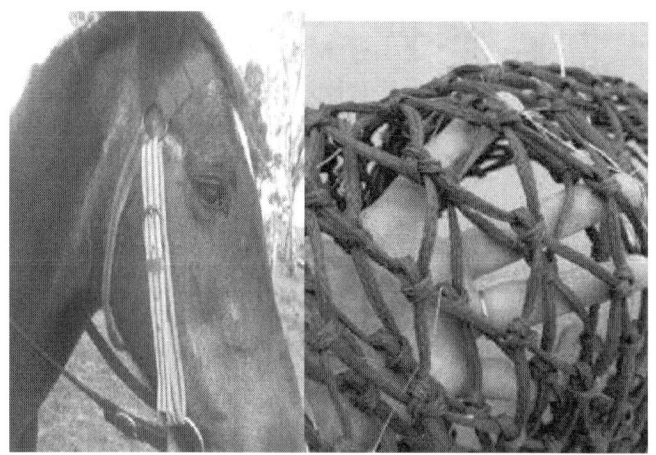

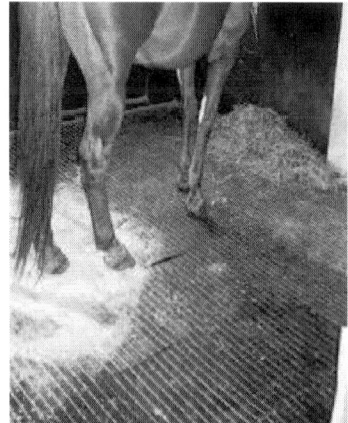

**Fig. 5.16(a)**    Agro Textiles for Animal Husbandry

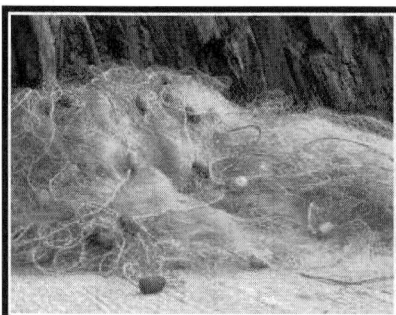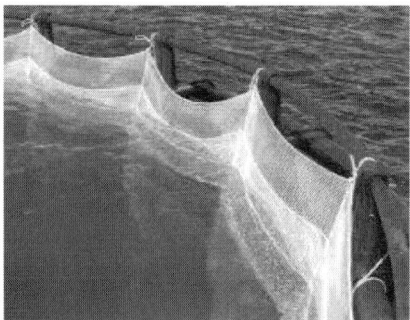

**Fig. 5.16(b)** Agro Textiles for Aquaculture

## 5.4 Agro Textiles for Agro-Engineering-Related Applications

### 5.4.1 Greenhouse

Woven greenhouse shade nets in different colours and weaves prepared from polyethylene and polypropylene yarns have been found to be comparable in cost and performance with commercially available knitted shade nets.

The classical definition of a greenhouse or a poly house is 'a framed structure covered with transparent or translucent material and large enough to grow crops under partial or fully controlled environmental conditions to get maximum productivity and quality produce'. Woven greenhouse shade nets, in a range of light-to-heavy and wide-width fabrics, can be made on projectile weaving machines. Nets with a mesh width of 1.8–40 mm can be produced (Fig. 5.17).

Greenhouse shade nets are used to protect crops and plants from adverse weather conditions, animals and pests, besides providing suitable conditions for plant growth. Some of the essential performance properties required of greenhouse shade nets are as follows.

(i) *Resistance to solar radiation and weathering*: Energy from the sun is transmitted through the shade net to the plant, which is then used in photosynthetic processes. The intensity of the photosynthetically active radiations (400–700 nm) directly influences plant growth. Other non-visible radiations include UV, IR and far IR wave bands. The greenhouse shade net material must withstand weathering effects of these radiations. However, no single material is resistant to all radiations – polypropylene and polyester are more resistant to UV radiations than polyethylene. Polyethylene is resistant to radiations in the visible region.

(ii) *Abrasion resistance*: The use of greenhouse shade nets in outdoor conditions also requires them to be resistant to abrasion. The abrasion to which a shade net is subjected may be of the material itself (material-to-material) or of stray animals. Abrasion of the shade net would result in holes though which animals and pests could enter the structure and harm the crops. Good abrasion resistance is required of greenhouse shade nets.

(iii) *Tensile strength*: The tensile strength of shade nets can be a deciding factor of its long-term durability and service life. Hence good tensile strength is necessary parameter for shade nets.

**Fig. 5.17**   Greenhouses

## 5.4.2    Agro Bags

Agro bags are grow bags which are co-extruded, two or three layers (multilayer) polythene, in sheets, roll form and bag form, for example, UV-treated bags (UV treatment) – 6 months to 5 years, with guarantee certification for UV stability and exposure periods in greenhouse conditions, hydroponic cultivation or industrial growing conditions. The plastic bag contains sufficient amount of growing medium and nutrients to enable plants such as tomato, bell pepper and strawberry to be grown to its full potential (Fig. 5.18).

**Fig. 5.18**   Agro Bags

### 5.4.3    Soil Covers

Soil cover refers to vegetation, including crops, and crop residues on the surface of the soil (Fig. 5.19).

A permanent year round soil cover is central to conservation agriculture. It is important for several reasons:

- It protects the soil from rain, sun and wind. It reduces soil erosion and protects the fertile topsoil, so preventing the silting of rivers and lakes. It stops the soil surface from sealing and reduces the amount of precious rainwater that runs off.

- It suppresses weeds by smothering their growth and reducing the number of weed seeds. This reduces the amount of work needed for weeding.

- It increases the soil fertility and the organic matter content of the soil.

- It increases soil moisture by allowing more water to sink into the ground and by reducing evaporation. Decomposing vegetation and the roots of cover crops improve the soil structure and make the clumps and lumps in the soil more stable – making it harder for rain to break them up and wash them away.

- Earthworms and other forms of life can prosper in the cover as well as in the soil.

- Soil cover stimulates the development of roots, which in turn improve the soil structure, allow more water to soak into the soil and reduce the amount that runs off.

**Fig. 5.19**   Soil Covers

## 5.4.4    Grass Reinforcement

Grass reinforcement mesh is a heavy duty thick slip resistant polyethylene plastic mesh grid for reinforcing and protecting grassed surfaces prone to wear, rutting and smearing which can result in a muddy surface incapable of withstanding vehicular or pedestrian traffic applications (Fig. 5.20).

Grass reinforcement grid provides:

- High level of grass reinforcement.
- Up to 8 tonnes per axle (imposed load)
- Ideal for permanent or temporary applications
- Fast and cost-effective installation compared to plastic paving grids
- No excavation or soil removal necessarily required
- Up to 97% improved slip resistance compared to the standard grass protection meshes

**Fig. 5.20**    Grass Reinforcement

## 5.4.5    Packaging Material

**Fig. 5.21**   Protective Bags

Cloth bags are usually strong enough to protect fruit from marauding possums and birds. Fruit bats may manage to suck the fruit through the bag. These bags are sturdy washable calico cloth with a drawstring. Cloth bags have the disadvantage of staying moist longer after rain, so they are a better choice in less humid climates or dry seasons. Size 200 mm × 300 mm. Product supplied as individual bags.

Fruit protection bags are easy to use for stone fruit including peaches, nectarines, plums and apricots. This horticultural waxed paper bag is specifically designed to attach around the lateral branch rather than the fruit stem.

Mesh sleeves are innovative products are made of sturdy UV-resistant fly screen. The sleeves are open at both ends and are available at different sizes. They are designed to slide along a branch, or over a large bunch, and protect fruit from fruit fly, birds and possums. Once in position, it can be tied closed with the attached long lasting 'brickies string' (Fig. 5.21).

Mesh bags are available at different sizes. The smaller bag suits any individual or small clustering fruit while the larger size would suit fruit with several pieces closer to the end of a branch (loquat, mango, lychee, etc.).

## 5.4.6    Vermicomposting Beds

Vermiculture/vermicomposting through earthworm is economic and easy technology to convert all biodegradable waste into best quality organic manure. All types of biodegradable materials like urban solid waste/rural solid wastes/food processing industries' wastes (agro waste) can be used to produce organic manure by adopting vermiculture/vermicomposting process.

It is the most eco-friendly solution to reduce the most talked about problem of global warming by reduction of use of chemical fertilizers and pesticides. Sustainable use of self-produced organic manure helps improvement of soil structure, porosity and overall fertility of the soil. It can also help in the development of wasteland by use of vermicompost manure. It improves soil structure, texture, aeration and water holding capacity and prevents soil erosion. Vermi wash, that is, the liquid generated due to vermicomposting is a very effective and organic pesticide (Fig. 5.22).

Vermi beds are made with HDPE woven beds for vermiculture, which is extensively used for vermicomposting. It is well-decomposed stable, fine granular by-product from composting process mediated through activities of earthworms. The HDPE woven and laminated vermi bed maintains required temperatures and moisture. Improves the nutrient characteristic of the compost, decrease the time required for composting process. Conventionally only a cemented pit or bamboo structures were used which are not convenient. Use of vermi bed gives 30–50 % higher production.

**Fig. 5.22**   Vermicomposting Beds

## 5.4.7    Backyard Fruit Netting

Every year thousands of animals are injured in inappropriate netting of back yard fruit trees or discarded netting. It entangles birds, lizards, snakes, bats and the occasional possum. The netting cuts their mouths to ribbons as they try to bite themselves free, and wraps so tightly around them that circulation is cut-off and tissue dies, days or even weeks later. The animals die of thirst, starvation, strangulation or outright pain and fear in the nets (Fig. 5.23).

Fruit saver is fitted fruit tree nets to protect fruit against fruit fly, birds, bats, possums and rats. They are box shaped with a long skirt that gathers around the trunk of the tree. They come in two sizes and are made from 2 mm woven mesh that gives a 15% shade factor.

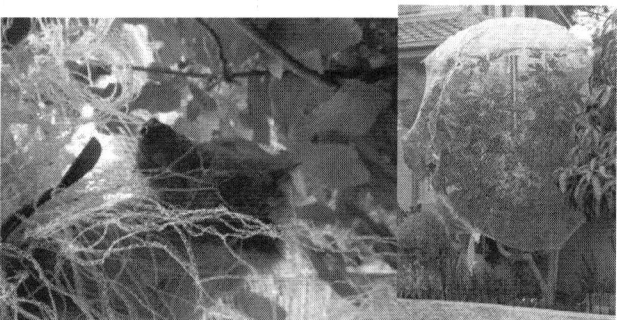

**Fig. 5.23** Backyard Fruit Netting

# Reference

1. Basu, S.K. Agricultural and horticultural applications of agro textiles, The Indian Textile Journal, Vol. 121, Issue 12, 2011, pp. 141–148.

2. Techtextil: Application Areas. Techtextil, Frankfurt: Trade Fair for Technical Textiles and Nonwovens, consulted 28 March 2015.

3. Subramaniam, V., Poongodi, G.R., Veena Sindhuja, V. Agro-textiles: production, properties & potential, The Indian Textile Journal, Vol. 119, Issue 7, 2009, pp. 73–77.

4. Kachru, R.P. Agro-Processing Industries in India – Growth, Status and Prospects. Indian Council of Agricultural Research, New Delhi, 2005, pp. 115–116.

5. NABARD. Report on Status and Potential of Village Agro Processing Industries. NABARD, Mumbai, 2005, pp. 17–26.

6. http://www.ifth.org

7. http://www.gcttg.com

8. http://www.textileroadmap.com

9. http://www.centexbel.be

10. http://www.technicaltextilesinfrance.com

11. http://textilelearner.blogspot.com

# 6

# Classification Based on Fabric Manufacturing Technologies

Textile fabrics are generally two-dimensional flexible materials made by interlacing of yarns or intermeshing of loops with the exception of non-wovens and braids. Fabric manufacturing is one of the four major stages (fibre production, yarn manufacturing, fabric manufacturing and textile chemical processing) of textile value chain. Most of the apparel fabrics are manufactured by weaving technology, though knitting is catching up fast especially in the sportswear segment. Natural fibres in general and cotton fibre in particular are the most popular raw material for woven fabrics intended for apparel use. Staple fibres are converted into spun yarns by the use of a series of machines in the yarn manufacturing section. Continuous filament yarns are texturized to impart spun yarn like bulk and appearance to them.

Textile fabrics are special materials as they are generally light weight, flexible (easy to bend, shear and twist), mouldable, permeable and strong (Fig. 6.1). The three major technologies of fabric manufacturing are as follows.

- Weaving
- Knitting
- Non-woven

There is use of synthetics as well as natural fibres in agro textiles, and the fibres used in agro textiles are nylon, polyester, polyethylene (PE), polyolefin, polypropylene (PP), jute and wool. Among all these fibres, polyolefin is extensively used, whereas among natural fibres, jute and wool are used, which not only serve the purpose but also after some year degrade and act as natural fertilizers. Several techniques offer specific advantages for particular product of fabric production which can be used to produce agro textiles.

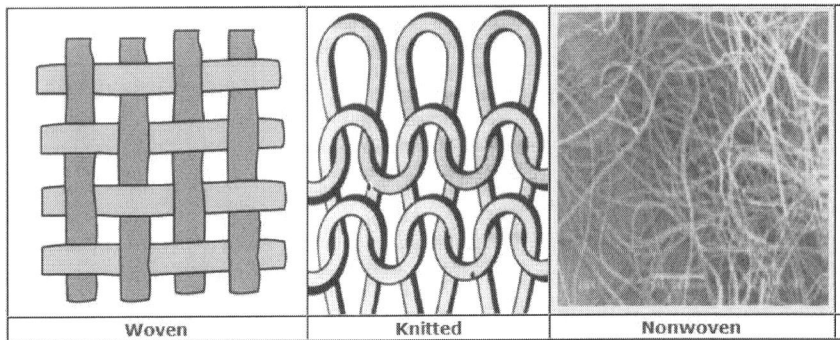

**Fig. 6.1**   Fabric Manufacturing Technologies

Fabric manufacturing may be preceded either by fibre production (in case of non-woven) or by yarn manufacturing (in case of weaving, knitting and braiding). Fabrics intended for apparel use must fulfil multidimensional quality requirements in terms of drape, handle, crease recovery, tear strength, air permeability, thermal resistance and moisture vapour permeability. However, looking at the unique properties and versatility of textile fabrics, they are now being used in various technical applications where the requirements are altogether different. Some examples are given in Table 6.1.

**Table 6.1**   Properties of Technical Fabrics

| Fabric type | Important properties/ parameters |
|---|---|
| Filter fabric | Pore size, pore size distribution |
| Body armour fabrics | Impact resistance, areal density, bending resistance |
| Fabrics as performs for composite | Tensile strength and tensile modulus |
| Knitted compression bandages | Stretchability, tensile modulus, creep |

## 6.1     Weaving Technology

Weaving is the most popular way of fabric manufacturing. It is primarily done by interlacing two orthogonal sets (warp and weft) of yarns in a regular and recurring pattern. Actual weaving process is preceded by yarn preparation processes namely winding, warping, sizing, drawing and denting.

Winding converts the smaller ring frame packages to bigger cheeses and cones while removing objectionable yarn faults. Pirn winding is performed to supply the weft yarns in shuttle looms. Warping is done with the objective to prepare a warper's beam which contains a large number of parallel ends in a double flanged beam. Sizing is the process of applying a protective coating on the warp yarns so that they can withstand repeated stresses, strains and flexing during the weaving process. Finally the fabric is manufactured on looms which perform several operations at proper sequence so that there is interlacement between warp and weft yarns and continuous fabric production.

## 6.1.1    Types of Looms

### 6.1.1.1    Handloom

This is mainly used in unorganized sector. Operations like shedding and picking are done by using manual power. This is one of the major sources of employment generation in rural areas.

### 6.1.1.2    Power Loom

It was designed by Edmund Cartwright in 1780s (during the industrial revolution). All the operations of the loom are automatic except the change of the pirn.

### 6.1.1.3    Automatic Loom

In this power loom, the exhausted pirn is replenished by the full one without stoppages. Under-pick system is a requirement for these looms.

### 6.1.1.4    Multiphase Loom

Multiple sheds can be formed simultaneously in this looms and thus productivity can be increased by a great extent. It has failed to gain commercial success.

### 6.1.1.5    Shuttle-less Loom

Weft is carried projectiles, rapiers or fluids in case of shuttle-less looms. The rate of production is much higher for these looms. Besides, the quality of the products is also better and the product range much broader compared to that of power looms. Most of the modern mills are equipped with different types of shuttle-less looms based on the product range.

### 6.1.1.6    Circular Loom

Tubular fabrics like hosepipes and sacks are manufactured by circular looms.

### 6.1.1.7    Narrow Loom

These looms are also known as needle looms and used to manufacture narrow width fabrics like tapes, webbings, ribbons and zipper tapes.

## 6.1.2    Primary Motions

Figure 6.2 shows some basic components of a loom. For fabric manufacturing through weaving, three primary motions are required namely shedding, picking and beat up.

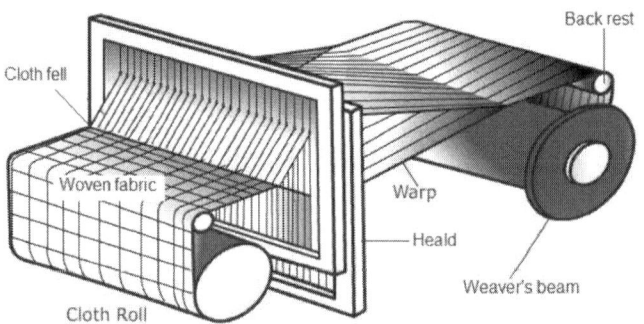

**Fig. 6.2**    Basic Loom Components

### 6.1.2.1    Shedding

It is the process by which the warp sheet is divided into two groups so that a clear passage is created for the weft yarn or weft carrying device to pass through it. One group of yarns (red yarns) either move in the upward direction or stay in the up position (if they are already in up position) (Fig. 6.3). Thus they form the top shed line. Another group of yarns (green yarns) either move in the downward direction or stay in the down position (if they are already in down position). Thus they form the bottom shed line.

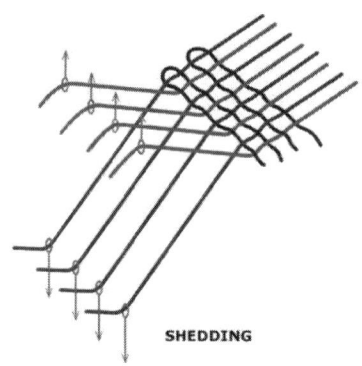

**Fig. 6.3**    Shedding

Except for jacquard shedding, warp yarns are not controlled individually during the shedding operation. Heald is a part that is related to the shedding mechanism. The heald shaft is made of wood or metal such as aluminium. It carries a number of heald wires through which the ends of the warp sheet pass. The heald shafts are also known as 'heald frames' or 'heald staves'. The number of heald shafts depends on the warp repeat of the weave. It is decided by the drafting plan of a weave (Fig. 6.4).

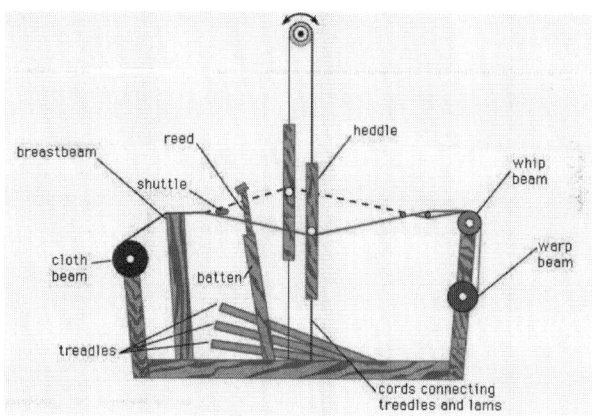

**Fig. 6.4**  Heald

### 6.1.2.2    Picking

The insertion of weft or weft carrying device (shuttle, projectile or rapier) through the shed is known as picking. Based on picking system, looms can be classified as follows.

- *Shuttle loom*: weft package is carried by the wooden shuttle
- *Projectile loom*: weft is carried by metallic or composite projectile
- *Air-jet loom*: weft is inserted by jet of compressed air
- *Water-jet loom*: weft is inserted by water jet
- *Rapier loom*: weft is inserted by flexible or rigid rapiers

With the exception of shuttle loom, weft is always inserted from only one side of the loom. The timing of picking is extremely important especially in case of shuttle loom. The shuttle should enter into the shed and leave the shed when the shed is sufficiently open (Fig. 6.5). Otherwise, the movement of the shuttle will be obstructed by the warp yarns. As a result, the warp yarns may break due to abrasion or the shuttle may get trapped in the shed which may cause damage to reed, shuttle and warp yarns.

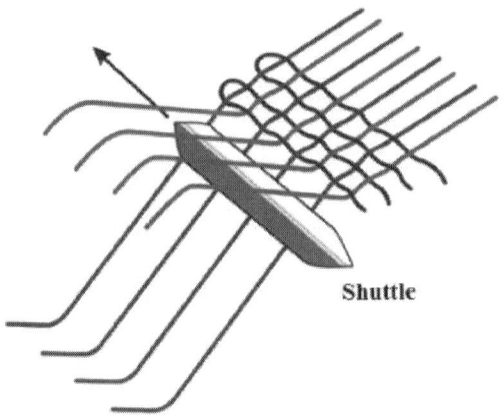

**Fig. 6.5** Picking

### 6.1.2.3 Beat up

Beat up is the action by which the newly inserted weft yarn is pushed up to the cloth fell (Fig. 6.6). Cloth fell is the boundary up to which the fabric has been woven. The loom component responsible for the beat up is called 'reed'. The reed, which is like a metallic comb, is carried by sley which sways forward and backward due to the crank-connecting rod mechanism. This is known as crank beat up. In modern looms, beat up is done by cam mechanism which is known as cam beat up. Generally, one beat up is done after the insertion of one pick.

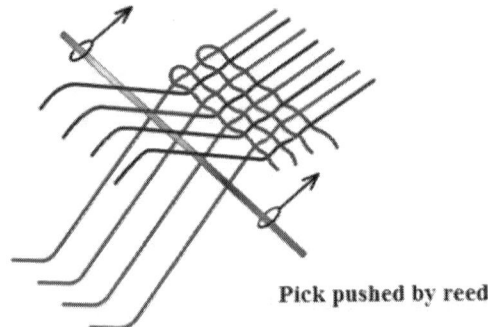

**Pick pushed by reed**

**Fig. 6.6** Beat Up

## 6.1.3  Secondary Motions

For uninterrupted manufacturing of fabrics, two additional secondary motions are required. These are take-up and let-off. Take-up motion winds the newly formed fabric on the cloth roller either continuously or intermittently after the beat up. The take-up speed also determines the picks/centimetre value in the fabric at loom state. As the take-up motion winds the newly formed fabric, tension in the warp sheet increases. To compensate this, the weaver beam is rotated by the let-off mechanism so that some new warp sheet is released.

## 6.1.4    Auxiliary Motions

Auxiliary motions are mainly related to the activation of stop motions in case of any malfunctioning like warp breakage, weft breakage or shuttle trapping within the shed. The major auxiliary motions are as follows:

* Warp stop motion (in case of warp breakage)
* Weft stop motion (in case of weft breakage)
* Warp protector motion (in case of shuttle trapping)

## 6.1.5    Use of Woven Fabrics in Agro Textiles: Polypropylene Woven Shade Cloth Fabrics

The PP is the most-used polymer for woven shade cloth fabrics. The resin is formulated with additives and pigments to provide resistance to sunlight and weathering. Woven shade cloth is constructed from woven PP. It is ideal for use on greenhouses and is typically 30% heavier than knitted PE shade cloth. It has the benefit of a longer lifespan of about 10–12 years. The drawback is that it will unravel if it is cut or a puncture occurs unlike knitted shade cloth and its uses are not as versatile as a knitted shade cloth.

Black pigmentation helps provide a high degree of sunlight resistance. Much of the shade cloth is made from monofilament yarn, although some film fibre yarns also are used. Wide-width fabrics minimize the amount of seaming needed for installation (Fig. 6.7).

**Fig. 6.7**   Leno/lock Woven Polypropylene Shade Material

### 6.1.6    Polyolefin Woven Shade Cloth Fabrics

Woven fabrics made with PP monofilament are also used as side curtains for wooden poultry houses. The curtains provide ventilation, yet protect birds from foul weather. Some fabrics are extrusion coated and then run through spiked rollers to improve their porosity. Extrusion-coated PP fabrics are used for hog houses in some areas of the country. Farmers use white woven PP fabric for barn wall siding as a way to add ventilation and reduce building costs.

Insect screening for greenhouses is another important application for woven agro textiles. These fabrics are made from fine-denier PE or PP monofilament, typically with a size of 0.018 inch (Fig. 6.8).

**Fig. 6.8**    Polyolefin Shade Material

## 6.2    Knitting Technology

Knitting is a method by which yarn is manipulated to create a textile or fabric. Knitting creates multiple loops of yarn, called stitches, in a line or tube. Knitting has multiple active stitches on the needle at one time. Knitted fabrics are known to have existed even before 256 AD, evidenced by samples found in Syria and Egypt. This art of producing fabrics was introduced to Europe by the Arabs. The first knitting machine was invented in 1589 – incidentally 200 years prior to French revolution – by Rev. William Lee of Nottingham.

The resultant fabric, a matrix of rows and columns of loops, is formed by creating a single element in each complete cycle of operation. Hence, if a fabric needs to have 100 loops in each row, then 100 cycles of operation would be needed to produce one row. Subsequently, loops of the row just completed would be transferred one after the other to another pin and in the process new elements are generated for the next row. The machine invented by Lee could however generate one complete row in each cycle of operation.

Like weaving, knitting is a technique for producing a two-dimensional fabric made from a one-dimensional yarn or thread. In weaving, threads are always straight, running parallel either lengthwise (warp threads) or crosswise (weft threads). By contrast, the yarn in knitted fabrics follows a meandering path (a *course*), forming symmetric loops (also called bights) symmetrically above and below the mean path of the yarn. These meandering loops can be easily stretched in different directions giving knit fabrics much more elasticity than woven fabrics. Depending on the yarn and knitting pattern, knitted garments can stretch as much as 500%. For this reason, knitting was initially developed for garments that must be elastic or stretch in response to the wearer's motions, such as socks and hosiery.

For comparison, woven garments stretch mainly along one or other of a related pair of directions that lie roughly diagonally between the warp and the weft, while contracting in the other direction of the pair (stretching and contracting with the *bias*), and are not very elastic, unless they are woven from stretchable material such as spandex (Fig. 6.9). Knitted garments are often more form fitting than woven garments, since their elasticity allows them to contour to the body's outline more closely; by contrast, curvature is introduced into most woven garments only with sewn darts, flares, gussets and gores, the seams of which lower the elasticity of the woven fabric still further. Extra curvature can be introduced into knitted garments without seams, as in the heel of a sock; the effect of darts, flares, etc., can be obtained with short rows or by increasing or decreasing the number of stitches. Thread used in weaving is usually much finer than the yarn used in knitting, which can give the knitted fabric more bulk and less drape than a woven fabric.

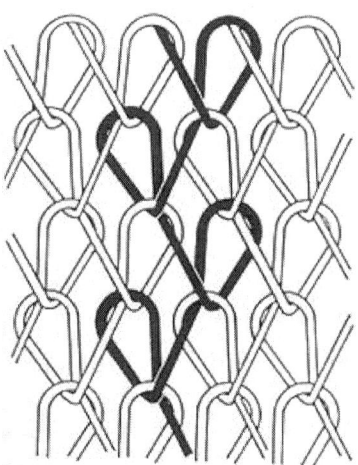

**Fig. 6.9**   Basic Pattern of Warp Knitting

## 6.2.1 Weft and Warp Knitting

There are two major varieties of knitting: weft knitting and warp knitting. In the more common weft knitting, the wales are perpendicular to the course of the yarn. In warp knitting, the wales and courses run roughly parallel (Fig. 6.9). In weft knitting, the entire fabric may be produced from a single yarn, by adding stitches to each wale in turn, moving across the fabric as in a raster scan. By contrast, in warp knitting, one yarn is required for every wale. Since a typical piece of knitted fabric may have hundreds of wales, warp knitting is typically done by machine, whereas weft knitting is done by both hand and machine. Warp-knitted fabrics such as tricot and milanese are resistant to runs, and are commonly used in lingerie (Fig. 6.10).

**Fig. 6.10** A Modern Knitting Machine in The Process of Weft Knitting

## 6.2.3 Knitting Technologies in Manufacture of Technical Textiles

Various kinds of warp-knitted fabrics play the most important role among knitted technical textiles. These fabrics are made on tricot, raschel, crochet and knit braiding machines. Products of these machines can be used in themselves, like nets or bandages, but also as reinforcement materials in composites or backing materials for laminated or coated fabrics.

However, besides warp knitting technology, important products are made also on weft knitting, mainly on circular knitting machines but V-bed flat knitting must not be neglected either.

### 6.2.3.1 Nets

Application field of nets is extremely wide, for example, agriculture, fishing, packaging, transport, sports, shading technology, construction, healthcare, surgery, safety technology and military, etc. Many of these nets are made by raschel or crocheting technology the great advantage of which is that they do not contain knots (Fig. 6.11). This makes the nets easier to handle because the layers do not tangle up and there are no knots that could harm the good packed into the net. Warp-knitted nets – both flat and tubular ones – can be produced with very high productivity.

**Fig. 6.11** Raschel Net Without Knots

Raschel net does not contain knots. Materials used for net manufacture are very different, depending on the end-use. Spun yarns or filament yarns and narrow plastic tapes are commonly used for this purpose. Elastic nets are made with using of elastane yarns. Width of flat nets knitted on raschel or tricot machines may reach as well 5–6 m, while to make narrower variants (up to 100–120 cm width), crochet machines are also available. Raschel machines with two needle bars are able to produce wide tubular net fabrics. To manufacture tubular nets of smaller diameters (from 1 or 2 cm to about 20 cm), knit breading machines can be used very effectively but their final diameter can be extended in the practice if they contain elastane yarns.

## 6.2.4 Knitted Fabrics with Orientated Behaviours

Knitted fabrics with orientated behaviours are made usually with lots of yarns laid lengthwise, crosswise and/or diagonally into the fabric. Their keeping together is performed by warp knitted loops. Aim of these structures is mainly to reduce the stretch and/or to increase the forth of the fabric in one or more directions. If this effect is realized only in one direction (lengthwise or cross-wise), the fabric is called 'unidirectional' or 'monoaxial'. If this behaviour asserts itself in both directions, the fabric is called 'biaxial'. 'Multiaxial' or 'multidirectional' fabrics have almost the same behaviour in every direction (Fig: 6.12).

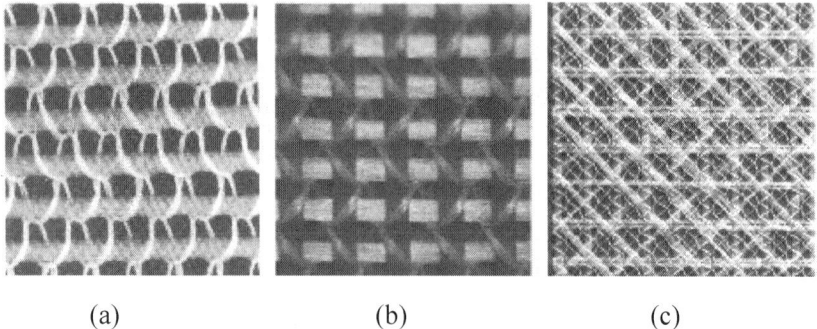

(a)                    (b)                    (c)

**Fig. 6.12**   Knitted Fabrics with Orientated Behaviours: (a) Monoaxial, (b) Biaxial and (c) Multiaxal

# 6.3    Non-Woven

There are many techniques to produce non-woven fabrics. Spun bonding and needle punch techniques are mainly used for the production of non-woven agro textiles. The spun-bonded fabric has high and constant tensile strength in all directions. It has also good tearing strength. Needle-punched fabric plant bags provide advantages over conventional fired clay pots. All natural fibres offer an added advantage of that the container decomposes after being planted in the ground.

Non-woven fabrics are products made of parallel laid, cross laid or randomly laid webs bonded with application of adhesive or thermoplastic fibres under application of heat and pressure. In other words, non-woven fabric can be simply defined as a fabric those can be produced by a variety of processes other than weaving and knitting.

## 6.3.1    Properties of Non-woven

Fabric properties of non-wovens range from crisp to that soft-to-the-touch to harsh, and from impossible-to-tear to extremely weak. This leads to a wide range of end products such as nappies, filters, teabags, geotextiles, etc., some of which are durable and others are disposable. The first stage in the manu-facturing process of non-woven fabrics is 'production of web' and another is 'bonding of web by using several methods'. Some of those (binding methods) are felting, adhesive bonding, thermal bonding, stitch bonding, needle punching, hydro-entanglement and spin laying.

The non-woven fabric properties depend on the following particulars to a great extent:

1. Choice of fibres
2. Technology used
3. Bonding process
4. Bonding agent

## 6.3.2    Techniques Used for Non-woven Production

There are many techniques to produce non-woven fabrics.

1. Needle punching
2. Spun bonding
3. Thermal bonding

### 6.3.2.1    Needle Punching Process

Worldwide, the needle punching industry enjoys one of the greatest successes of any textile-related process. The needle punching industry around the world is a very exciting and diverse trade involving either natural or both natural and synthetic fibres.

The needle punch process is illustrated in Fig. 6.13. Needle-punched non-wovens are created by mechanically orienting and interlocking the fibres of a spun bonded or carded web. This mechanical interlocking is achieved with thousands of barbed felting needles repeatedly passing into and out of the web.

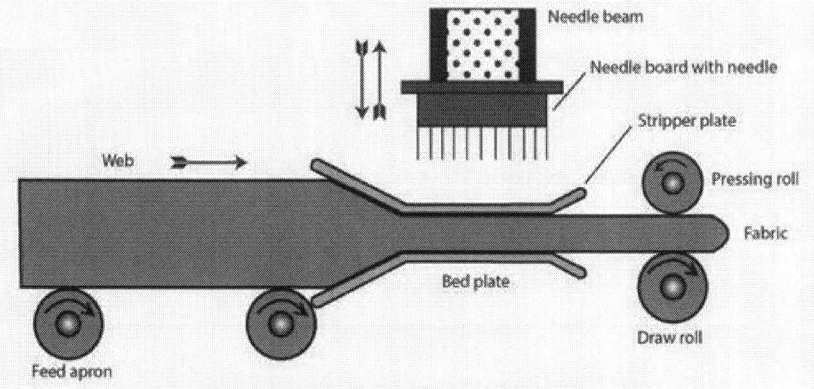

**Fig. 6.13**  Needle Punching Process

The major components of the needle loom and brief description of each are discussed in the following sections.

## The Needle Loom

- The needle board: The needle board is the base unit into which the needles are inserted and held. The needle board then fits into the needle beam that holds the needle board into place.

- The feed roll and exit roll: These are typically driven rolls and they facilitate the web motion as it passes through the needle loom.

- The bedplate and stripper plate: The web passes through two plates, a bedplate on the bottom and a stripper plate on the top. Corresponding holes are located in each plate and it is through these holes the needles pass in and out. The bedplate is the surface the fabric passes over which the web passes through the loom. The needles carry bundles of fibre through the bedplate holes. The stripper plate does what the name implies; it strips the fibres from the needle so the material can advance through the needle loom.

## The Felting Needle

The correct felting needle can make or break the needle-punched product. The proper selection of gauge, barb, point type and blade shape (pinch blade, star blade, conical) can often give the needle puncher the added edge needed in this competitive industry.

The gauge of the needles is defined as the number of needles that can be fitted in a square inch area. Thus, finer the needles, higher the gauge of the needles. Coarse fibres and crude products use the lower gauge needles, and fine fibres and delicate fibres use the higher gauge needles. For example, a sisal fibre product may use a 12–16 gauge needle and fine synthetics may use 25–40 gauge needle.

The major components of the basic felting needle are as follows:

- Crank: The crank is the 90° bend on the top of the needle. It seats the needle when inserted into the needle board.

- Shank: The shank is the thickest part of the needle. It fits directly in the needle board itself.

- Intermediate blade: The intermediate blade is put on fine gauge needles to make them more flexible and somewhat easier to put inside the needle board. This is typically put on 32 gauge needles and finer.

- Blade: The blade is the working part of the needle. The blade is what passes into the web and is where the all-important barbs are placed.

- Barbs: The barbs are the most important part of the needle. It is the barb that carries and interlocks the fibres the shape, and size of the barbs can dramatically affect the needled product.
- Point: The point is the very tip of the needle. It is important that the point is of correct proportion and design to ensure minimal needle breakage and maximize surface appearance.

As the needle loom beam moves up and down the blades of the needles penetrate the fibre batting. Barbs on the blade of the needle pick up fibres on the downward movement and carry the fibres the depth of the penetration. The draw roll pulls the batt through the needle loom as the needles reorient the fibres from a predominately horizontal to almost a vertical position. The more the needles penetrate the web, the more dense and strong the web. Beyond some point, fibre damage results from excessive penetration.

**Types of Looms**

There are three basic types of needle looms in the needle punching industry. They are as follows:

- Felting loom
- Structuring loom
- Random velour loom

The felting looms are the type just described. These needle looms may have one to four needle boards and needles from the top, bottom or top and bottom. The primary function of this type of loom is to do interlocking of fibres resulting in a flat, one-dimension fabric. The types of products made with this process and needle loom are diverse and multifaceted. They exist in variety of industrial products, geotextiles, automotives, interlinings, home furnishings, etc.

Structuring looms use what are called fork needles. Instead of carrying fibres into bedplate hole, the fork needles carry fibre tufts into lamella bars that extend from the entry to the exit of the needle loom. These fork needles carry large tufts of fibres into parallel lamella bars. These bars carry the tuft of fibre from the entry to the exit side of the loom. Depending on the orientation of the fork needle, a rib or velour surface is introduced. The most popular products made with structuring looms include home and commercial carpets and floor mats, automotive rib and velour products, wall covering and marine products.

Random velour looms are the newest type of needle looms, having only been available since the mid-1980s. The random velour looms are used to produce velour surfaces. Unlike the structuring looms, the velour products produced by this loom are completely isotropic. It is almost impossible to distinguish the cross direction from the machine direction.

Unique to this type of needle loom is the bristle-brush, bedplate system. Special crown type needles or fork needles are used in this loom design. The needles push fibres into a moving brush bedplate. The fibres are carried in this brush from the entry to the exit of the loom with zero draft. This allows for the completely non-linear look, perfect for moulded products. Random velour type products have been very popular in the European and Japanese automotive industry. While almost all the US automotive producers have the random velour machine, this type of product has yet to become popular in this country. The most popular products made with this type of needle loom are almost all centred around the automotive industry.

## Applications of Needle Punching Non-wovens

*1. Non-wovens as crop covers or row covers*

Non-wovens are replacing some of the straw, glass and plastic films that have been used for many years to protect crops from freezing. They now are used to accelerate plant growth early in the season.

Lightweight cotton fabrics were used for many years to prevent newly planted seeds from being washed away. Stabilized spun bond PP fabrics that weigh in the range of 0.3 to 1 ounce per square yard are replacing cotton fabrics for this use and also may be used as a crop cover.

Most vegetables respond well to the use of floating row covers of non-woven materials. Generally, covers are used from a week before the frost-free date until flowers appear 4–6 weeks later. Covers are stored and sometimes put back in the fall to protect the fruit or vegetables from an early frost. Early yields bring more dollars per pound for the farmer, and higher yields result in more total dollars.

Strawberries are a crop for which early and total yields have been enhanced by floating row covers. They can be protected with row covers down from 22°F to 24°F, depending on wind speed and freeze duration. Tree seedlings and nursery stock also benefit from the use of row covers, which prevent the heaving of seedlings by hoar frost as well.

*2. Non-wovens as Landscaping Fabrics*

The PP spun bond non-wovens are the dominant products used for landscaping fabrics because of their durability and relatively low cost. Some of the major landscaping uses for the fabrics are as follows:

- Soil retention and weed control for landscaped areas and gardens
- Soil retention for retaining walls made of timber
- Weed control under decks
- Brick walkway and patio support
- Planter and pot drainage
- Linings for interceptor trench drains
- Protection for newly seeded areas

One successful specialty agro textile is Biobarrier® fabric for root control, manufactured by Reemay Inc. (Hickory, TN, USA) – a member of BBA Fiberweb™. This product combines a spun bond PP fabric with a time-released herbicide, creating an in-soil herbicidal barrier that can block roots without harming plants. The system is guaranteed for 15 years. The pellets that are adhered to the spun bond are impregnated with the herbicide trifluralin. The herbicide is mixed with carbon black and PE and moulded into pellets. The specific formulation selected controls the rate of release of the herbicide.

Metalized woven and non-woven agro textile products are finding applications in greenhouses, plant nurseries and orchards. These fabrics may be used in greenhouses to prevent heat loss. Installations of these fabrics also can be designed to protect plants from excessive solar radiation.

Diversified Fabrics Inc. (Kings Mountain, NC, USA) uses its Reflec-Tex® process for metallizing non-woven and woven fabrics for agro textiles and other applications. F.J. Broadwell (President of Diversified Fabrics Inc.) reports the company is metallizing these fabrics for use in greenhouses and orchards. A specific application he cites involves using the metallized woven fabrics under apple and peach orchards. The reflected sunlight from the fabric hastens fruit ripening and provides fruit with more uniform colour. The first pickings are larger, and often a fewer number of pickings are required. The fabrics also minimize weed growth in the orchard.

### 6.3.2.2    Spun Bonding

In the spun bond technology, usually a thermoplastic fibre forming polymer is extruded to form fine filament fibres of around 15–35 μm diameter. The filaments are attenuated collected on a conveyor belt in the form of a web. The filaments in web are then bonded to make spun bond non-woven fabric.

### Raw Materials

Spun bond technology uses preferably thermoplastic polymers with high molecular weight and broad molecular weight distribution such as PP and polyester (polyethylene terephthalate – PET). To a small extent, other polyolefins such as high-density and low-density PE as well as a variety of polyamides (PAs), mainly PA 6 and PA 6.6, are found. Out of these polymers, PP is mostly used primarily due to its low price and advantageous properties such as low density, chemical resistance and hydrophobicity, sufficient or even better strength. The fibre grade PP (mainly isotactic) is the principal type of PP which is used in spun bond technology. The important raw material parameters for PP to be a suitable candidate for spun bond technology are melt flow index of about 20–40 g/10 min and polydispersity ratio $(M_w/M_n)$ of around 3.5–7.

The molecular weight can be around 180,000. On the contrary, the important raw material parameters for polyester are intrinsic viscosity of about 0.64, low share in COOH groups, high crystallinity and low water content (as low as 0.004%). Spun bond non-wovens are exclusively made from crystalline polyester. Crystallinity influences pre-drying and extrudability as well as filament drawing orientation, which is basic to make products that meet the requirements and that are of proper strength. Pre-drying is inevitable as PET at thermal strain is subject to hydrolytic degradation when extruded. In addition, low water content avoids air pockets in the melt that might cause filament breakage. Frequently, requirements can only be met by means of polymer modification. Except for the mechanical properties, ultraviolet resistance and flame retardancy are important with technical applications.

Nowadays, the bicomponents are found in spun bond fabrics. The cross section of these bicomponent filaments has at least two different polymer components. Figure 4.1 shows different geometry of cross sections of the bicomponent filaments. Sometimes the bicomponent filaments are splitted or fibrillated into microfibres by means of hydroentangling energy. The resulting fabrics are extremely soft, particularly after finishing, and have therefore been considered for use in clothing, hygiene and medical dressing components. In addition, bicomponent fibres with eccentric sheath core arrangement are used to develop crimp in spunlaid fabrics by differential thermal shrinkage of the two polymer components.

**Process Sequence**

Figure 6.14 displays a schematic diagram of spun bond machine. The spun bond technology, in its simplest form, consists of four processes namely spinning, drawing, web formation and web bonding. The spinning process largely corresponds to the manufacture of synthetic fibre materials by melt-spinning process. In the drawing process, the filaments are drawn in a tensionally locked way. The web formation process forms a non-woven web. Web bonding is generally possible by means of the web bonding processes discussed earlier. The bonding process includes mainly thermal calender bonding. Mechanical bonding and chemical bonding of spunlaid webs are also reported. The sequence of processes is as follows: polymer preparation → polymer feeding, melting, transportation and filtration → extrusion → quenching → drawing → lay down → bonding and winding.

The first step to spun bond technology involves preparation of polymer. It involves sufficient drying of the polymer pellets or granules and adequate addition of stabilizers/additives. The drying of the polymer is carried out particularly for polyester and PAs as they are relatively high hygroscopic than PP. The stabilizers are often added to impart melt stability to the polymers. Then, the polymer pellets or granules are fed to an extruder hopper by gravity feeding. The pellets are then supplied to an extruder screw, which rotates within the heated. As the pellets are conveyed forward along the hot walls of the barrel between the flights of the screw, the polymer moves along the barrel, it melts due to the heat and friction of the viscous flow and the mechanical action between the screw and barrel. The screw is divided into feed, transition and metering zones. The feed zone preheats the polymer pellets in a deep screw channel and conveys them into the transition zone. The transition zone has a decreasing depth channel to compress and homogenize the melting plastic. The melted polymer is discharged to the metering zone, which serves to generate maximum pressure for pumping the molten polymer. The pressure of the molten polymer is highest at this point and is controlled by the breaker plate with a screen pack placed near the screw discharge. The screen pack and breaker plate also filter out dirt and unmelted polymer lumps. The pressurized molten polymer is then conveyed to the metering pump.

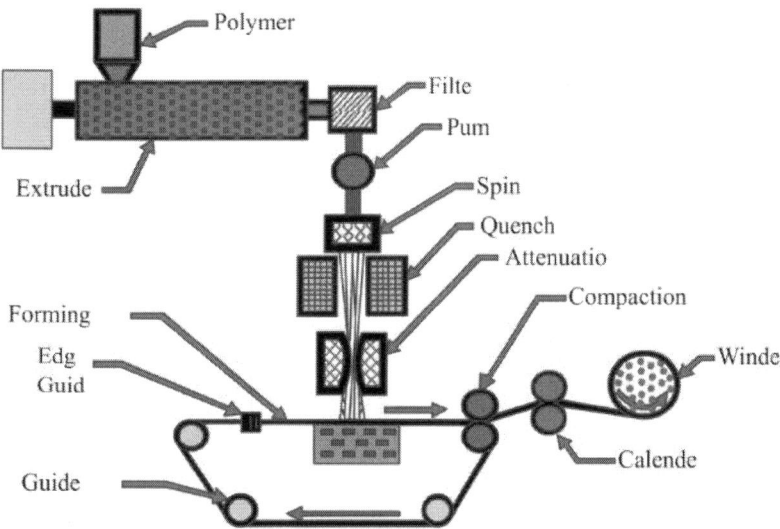

**Fig. 6.14**   Schematic Diagram of Spun Bond Machine

### 6.2.3.3    Thermal Bonding

It is known that the fibres in the webs can be bonded thermally to have sufficient resistance to mechanical deformation. The basic concept of thermal bonding was introduced by Reed in 1942. He described a process in which a web consisting of thermoplastic and non-thermoplastic fibres was made and then heated to the melting or softening temperature of the constituent thermoplastic fibres followed by cooling or solidifying the bonding area. Since then many developments have been made in thermal bonding processes. Today the thermal bonding processes include calender bonding, through air bonding, infrared bonding and ultrasonic bonding. Thermal bonding requires a thermoplastic component to be present in the web in the form of homofil fibre, powder, film, hot melt or as a part (sheath) of bicomponent fibre. The thermoplastic component becomes viscous under the application of thermal energy. The polymer flows to fibre-to-fibre crossover points where bonding regions are formed. The bonding regions are fixed by subsequent cooling. The thermal bonding process is environmental friendly, as no latex binder is required. The thermal bonding process consumes less energy compared to foam bonding or hydroentanglement bonding.

## Principle of Thermal Bonding

The formation of a bond during thermal bonding follows in sequence through three critical steps:

1. Heating the web to partially melt the crystalline region

2. Repetition of the newly released chain segments across the fibre–fibre interface

3. Subsequent cooling of the web to re-solidify it and to trap the chain segments that diffused across the fibre–fibre interface

The time scales for these processes closely match commercial practice.

The formation of a bond requires partial melting of the crystals to permit chain relaxation and diffusion. If, during bonding, the temperatures are too low or if the roll speeds are too high, the polymer in the mid-plane of the web does not reach a high enough temperature to release a sufficient number of chains or long enough chain segments from the crystalline regions. Thus, there will be very few chains spanning the fibre–fibre interface, the bond itself will be weak, and the bonds can be easily pulled out or ruptured under load, as observed. Under-bonding occurs when there is an insufficient number of chain ends in the molten state at the interface between the two crossing fibres or there is insufficient time for them to diffuse across the interface to entangle with chains in the other fibre. Over-bonding occurs when melting occurs and many chains have diffused across the interface and a solid, strong bond has been formed. If the web reaches a sufficient temperature, many chains or chain segments are released from the crystal, repeat across the fibre–fibre interface and form a strong bond. The fibres within the bond spot have lost their orientation and their strength. At the same time, the polymer chains within the fibres located in the vicinity of the bond also lose some of their molecular orientation (and strength) at the fibre-bond interface.

In well-bonded webs, failure occurs at the bond periphery because the bridging fibres are weak in the region adjacent to the bond, but strong elsewhere. If the bridging fibres have the same strength over their entire length, including the region at the bond periphery, better load sharing would lead to a stronger web.

### Raw Materials

The thermal bonding processes utilize either thermoplastic fibres alone or blends containing fibres that are not intended to soften or flow on heating. The non-binder fibre components may be referred to as the base fibres or sometimes carrier fibres. Commercially, a variety of base fibres are used. The binder fibre component normally ranges from 5% to 50% on weight of the fibre depending on the targeted properties of the final product made thereupon.

## References

1. Adanur, S. Hand Book of Industrial Textiles. Technomic Publishing Co., Inc., Lancaster, PA, USA, 1995.

2. Adanur, S. Hand Book of Weaving. Technomic Publishing Co., Inc., Lancaster, PA, USA, 2001.

3. Koerner, R. Designing With Geo Synthetics. Prentice Hall, Upper Saddle River, NJ, USA, 1994.

4. Sambhasivarao, K. Handbook for Agrotextiles.

5. Jaiswal, H., Barhanpurkar, S., Chandak, S., Kabra, N. Textile at Agriculture Application.

6. Agarwal, S.K. Application of Textile in Agriculture, College of Engineering & Technology, Akola, Maharashtra.

7. Application of technical textiles, agro textile and home textile, The Indian Textile Journal.

8. www.textilelearner.blogspot.com

9. www.ijarse.com

10. www.indiantextilejournal.com

11. www.technicaltextile.net

12. www.fiber2fashion

13. textilelearner.blogspot.com.tr/2014/04/applications-of-agro-textiles.html

14. ttps://vibrantgujarat.com/.../images/.../agrotech-textile-application-in-agriculture.pdf

15. www.indiantextilejournal.com/articles/FAdetails.asp?id=1999

16. www.fibre2fashion.com › Knowledge › Article

17. www.authorstream.com/.../MAYURTIL-2265228-agro-textile-11-12-september/

# 7
# Classification Based on Products

The selection of agro textile product is depends on crop needs. Selection of the agro textiles is also greatly influenced by the geographical location. Some of the applications of agro textiles are also based on the products. With the continuous increase in population worldwide, stress on agricultural crops has increased. To keep grains, vegetables and flowers, it is necessary to increase the yield and quality of agro products. Today, agriculture, horticulture area has realized the need of tomorrow and opting for various technologies to get higher overall yield, quality and tasty agro products. Adopting the hi-tech farming technique, where textile structures are used, could enhance quality and overall yield of agro products. Coir is a biodegradable organic fibre and hardest among other natural fibres. It is much more advantageous in different application for agricultural textiles. Coir is used commercially for the manufacture of wide range of products for varies end-use applications.

Textile structures in various forms are used in shade house/poly house, greenhouse and also in open fields to control environmental factors such as temperature, water and humidity. Agro products avoid damage from wind, rain and birds. Agro textiles such as sunscreen, bird net windshield, mulch mat, hail protection net, harvesting net, etc., are also used.

## 7.1 Shade Nets

Shade nets are nets made of polyethylene or polypropylene thread with specialized ultraviolet (UV) treatment having different shade percentages. These nets provide a partially controlled environment by primarily reducing light intensity and effective heat during day time to crops grown under it. This enables lengthening of the cultivation seasons and well as off-season cultivation depending on the conditions and type of crop. Shade nets are typically used in structures known as shade net houses which are frame structures made of materials such as galvanized iron pipes, angle iron, wood or bamboo which are then covered with shade nets to provide the benefits listed above. Each plant has its individual and unique requirements for sunlight and shade under which it flourishes at its best. To create optimum climatic conditions, selection of the correct percentage of shade factor plays an important role to enhance plant's productivity to its optimum.

If the conditions (factors) that the plant needs are improved, the rate of photosynthesis will increase. The maximum rate of photosynthesis will be constrained by a limiting factor. This factor will prevent the rate of photosynthesis from rising above a certain level even if conditions needed for photosynthesis are improved. If factors such as intensity of light, temperature, etc., are increased beyond limiting factor, then it will have an adverse effect on growth and yield. Thus, using suitable shade net will enable to modify or create an optimum protected environmental condition, which will result in enhanced quality and quantity of yield.

(1) Helps in cultivation of flower plants, foliage plants, medicinal and aromatic plants, vegetables and spices
(2) Used for raising nurseries of fruits and vegetable
(3) Helps to enhance yield during summer season
(4) Protects against pest attack
(5) Protects crops from natural weather disturbances such as wind, rain, hail, frost, snow, bird and insects
(6) Used in production of graft saplings and reducing its mortality during hot summer days
(7) Used for hardening tissue culture plants
(8) Helps in quality drying of various agro products
(9) Helps in creating favourable microenvironment for production of vermicompost.

Overall 35% were used for roses, strawberry, gooseberry, tomatoes, cucumber and fruit bearing plants, 50% for plants that grow under partial light like general pot and foliage plants and cut greens, orchids, *Anthuriums*, ginger, 75% for plants that grow under extreme low light such as indoor plants, certain orchids, plantation crops, tea, coffee and useful in summer to reduce the light level, 90% for usage in cattle sheds, poultry houses, construction scaffolding and vehicular shades (Fig. 7.1).

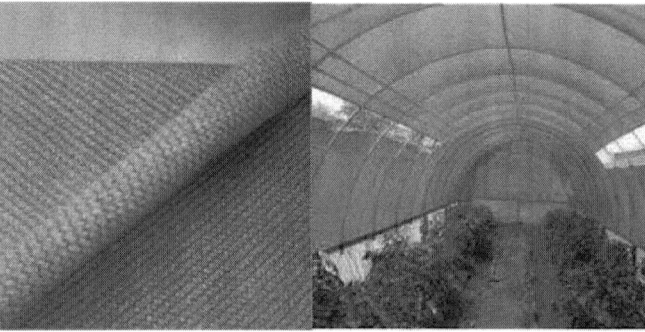

**Fig. 7.1**  Shade Nets

## 7.2    Agriculture Nets

Agriculture nets for agriculture lands, which are being used in different applications across any type of agriculture lands. These nets are made from high-grade quality raw materials to give long lasting service. These agriculture shade net seedlings are better grown under shade net for faster growth. These nets are also used to cover nurseries.

Agro shade net, which proves to be the best UV rays stabilizer. These shades provide paramount protection to plants kept inside the shade. The sheds also help in preventing plants from bad weather conditions (Fig. 7.2).

The shading nets fulfil the task of giving appropriate microclimate conditions to the plants. For instance, plants or shrubs from tropical environments usually grow under big trees and under a deep shade, mainly using the diffuse light that filters through the treetops. Besides, these plants require relatively high humidity and temperatures.

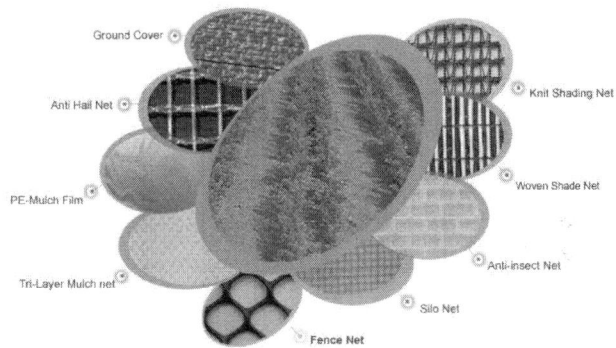

**Fig. 7.2**    Types of Shade Nets

## 7.3    Floriculture

Floriculture shade net is designed to provide protection from the sun to plants, crops and activity areas. The reduction in heat and UV protects sensitive plants, as well as reducing the effects of drying damage from wind. Shade cloth is also used for shade sails, temporary screens for building and construction and a variety of other applications.

Monofilament shade net provides protection against UV exposure, heat, wind, rain and hail. Its numerous uses include plant protection, scaffold cover and site screening, wind break protection and car protection. It is a high-quality knitted shade cloth manufactured using only virgin resins and the best UV stabilizers from Ciba Specialty Products that are available. Monofilament shade cloth comes in a range of colours, sizes, shades and UV constructions. Its numerous uses include plant protection, scaffold cover and site screening, wind break protection and car protection (Fig. 7.3).

**Fig. 7.3** Floriculture Shade Net

## 7.4 Scaffolding Nets

Monofilament/scaffolding nets are used on construction sites for scaffold enclosure, vision barriers, wind protection, pedestrian and traffic protection, trash containment, and for window and stair well safety. Workers on the edge of a roof should be protected with monofilament/scaffolding nets (Fig. 7.4).

Monofilament/scaffolding nets are suspended from cables that are supported by a bracket fastened. These same nets are used along the end walls as well as the side walls.

**Fig. 7.4** Scaffolding Nets

## 7.5     Shade Sail

These shades net give seedlings better grown under shade net for faster growth and its features such as high performance and long durability. Shade net gives extra safety from damaging UV rays, extreme sunlight, heat, cold, hail and winds. Originally shade fabric, like all fabrics in the outdoors, suffered from UV degradation. UV inhibitors are now added during the manufacture of shade cloth, and good shade cloth now generally comes with a multi-year UV degradation warranties. Shade cloth is a knitted fabric, and this is an important factor in design and manufacture shade sails (Fig. 7.5).

Successful shade sail design uses the inherent 'stretch' of the knitted fabric to create three-dimensional shapes. Fabrics other than shade cloth are used to make shade sails such as polyvinylchloride (PVC), a more expensive alternative, or canvas variations. The low cost of shade cloth and its ability to breathe makes it a prime choice for 'cool shade'.

**Fig.7.5**  Shade Sail

## 7.6     Agro Shade Net

About 35% density of agro shade net will block out 22% of light and 40% density of shade net will block out 40% of light. White shade net absorbs and reflects the white scattered light from the atmosphere. The interior of the structure is cooler than that covered by black shade net in hot weather conditions, as white shade cloth reflects the heat (Fig. 7.6).

**Fig. 7.6**  Agro Shade Net

## 7.7    Leno Bags

- High-density polyethylene (HDPE) mesh bag material recyclable, used for packing vegetable and fruit such as orange, potato, onion, carrot, garlic, cabbage, lemon, tomato, etc.
- Size 21×31 cm, the weight 8 g for packing about 3 kg, and 25×39 cm, the weight 10 g for packing about 5 kg
- 30×47 cm, the weight 12 g for packing 8 kg, and 40×60 cm, the weight 18–20 g for packing 10–15 kg
- 45×75 cm, the weight 25–27 g for packing 20–25 kg, and 50×80 cm, the weight 28–32 g for packing 25–40 kg. We accept customers design and also supply the tubular mesh bag for auto machine
- Colour red, green, orange, violet, yellow, white and any other colour
- Environmentally friendly and good appearance non-toxic economical (Fig. 7.7).

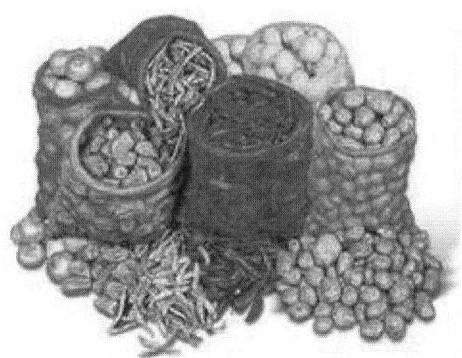

**Fig. 7.7**  Leno Bags

## 7.8    Garden Quilt

Garden quilt is a thicker version of our all-purpose fabric, consisting of polypropylene fibres that transmit 60% of available light. Garden quilt provides excellent frost protection (down to 24°F). The thick fabric is ideal for extending the growing season into early spring and late fall, or for insulating strawberries, herbs, perennials, small fruits and other tender landscape plants all winter long (Fig. 7.8).

**Fig. 7.8**    Garden Quilt

## 7.9    Frost Protection Fabric

Frost protection fabric creates a stable, favourable microclimate by capturing extra heat during the day and then slowing the loss of stored heat at night from the soil. As a result the fabric raises minimum temperatures, without suffocating, crushing or breaking plants the way plastic could. It is porous to allow air and water, plus the sunlight needed to flourish (Fig. 7.9).

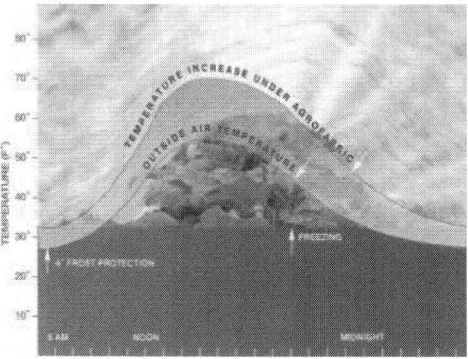

**Fig. 7.9**  Frost Protection Fabric

## 7.10    Wind Break And Shade

Windbreak and greenhouse shade netting offers great protection against potentially damaging gusts in the garden and also provides 50% shade when used in the greenhouse to protect plants from the scorching rays of the mid-summer sun (Fig. 7.10).

**Fig.7.10**  Frost Protection Fabric

## 7.11    Bird Control Net/Anti-Bird Netting

The three types of bird control netting, each with different characteristics and advantages, are discussed in the following sections

### 7.11.1    Extruded Netting

These polypropylene bird control nets provide an extremely cost-effective protection system for many kinds of plants and crops. The small mesh size prevents even the smallest of birds from attacking crops and plants. This mesh is considered to be environmentally friendly, since birds are unable to get into the crop and so will not be harmed by getting caught up within the net. The life expectancy of the net is one to two full seasons depending on conditions of use.

### 7.11.2    Knitted Netting

Knitted anti-bird nets are made from HDPE monofilament and offer long-term protection for all types of crops and plants against all birds and predators. A range of widths are available up to 12 m for larger areas. Life expectancy of the net is 4–6 years, depending on condition of use (Fig. 7.11).

### 7.11.3    Knotted Netting

The HDPE knotted anti-bird nets are used for many applications within horti-culture, agriculture and sports. This product is primarily used to guard against birds and is available in different grades and mesh sizes. The smaller the bird, the smaller the mesh. It is very easy to install, with a life expectancy of 5–7 years depending on condition of use.

**Fig. 7.11**   Anti-birds Netting

## 7.12    Anti-Hail Netting

Anti-hail net is made of polyethylene monofilament yarn with UV treatment. It is a narrow mesh open net and features high impact resistance. It can prevent heavy hailstorms and wind damaging crops, and it can reduce the economic loss. It is also used for preventing frost and reducing the heat loss. It protects effectively plants, orchard, garden, etc. (Fig. 7.12). The anti-hail net is used for protecting orchard, garden and botanical garden from heavy hail and wind damaging. It is also used for preventing birds biting the fruit and vegetables.

**Fig. 7.12**   Anti-Hail Netting

## 7.13    Nets For Covering Pallets

For safe transportation of fruit and vegetables to the market, the boxes are covered with large mesh nets and pallets to stop the boxes being turned upside down. This prevents damage of goods during transportation (Fig.7.13).

**Fig. 7.13**   Nets for Covering Pallets

## 7.14    Landscape Fabric

Landscape fabric is used for weed control, a central element in achieving low-maintenance landscaping. Effective weed control means a reduction in actual weeding or in herbicide use – both unsavory landscaping tasks. Thermally spun-bonded fabrics are said to be more effective than woven or needle-punched geotextiles in preventing fine roots and rhizomes from penetrating the fabric. While woven fabrics are very strong, they offer many spaces for weeds to penetrate. Needle-punched fabrics have loose threads of material where the plants can easily grow through. As for thermally spun-bonded fabrics, these have fibres fixed in place, keeping roots from penetrating (Fig. 7.14).

**Fig. 7.14**    Landscape Fabric

## 7.15    Wind Protection Fabric

Knitted windshield constructed from a commercial grade of UV stabilized yarn is often used to protect crops and structures from wind damage. Air exhaust deflection fabrics are otherwise most useful in odour control. These are often seen in swine and dairy house production. Wind protection fabric is available in a variety of aesthetically pleasing colours (Fig. 7.15).

**Fig. 7.15**   Wind Protection Fabrics

## 7.16    Poultry Curtains

Used more and more instead of traditional chicken wire to protect precious flocks of farm fowl, poultry curtains offer light control, thermal protection and ventilation control, even at sub-freezing temperatures (Fig. 7.16).

**Fig. 7.16**   Poultry Curtains

## 7.17    Textile Irrigation Systems

Optimal moisture management can be achieved using specialized soil covering materials. These fabrics, made from a multilayer high-performance textile, retain soil moisture, such as mulch, and insure a most favourable contact between precious water resources and the plant's roots, thus improving critical aspects of greenhouse tree nursery production. Textile irrigation systems currently being deployed generally have one layer that acts as a reservoir from which water is distributed equally and continuously throughout its surface, as well as a layer made from a light, low-density textile that prevents evaporation while transporting water to the pots through capillarity.

## 7.18    Flexible Silos

Natural products such as grain, animal feed and food place very special demands with regard to conservation on the farm. Flexible textile silos, made of active-breathing, dust-tight, very strong and durable polyester fabric, guarantee constant grain quality and healthy livestock. Contrary to solid silos, flexible silos will also limit condensation, which means that the formation of mould is effectively prevented. Finally, the galvanized steel frame and the jacket and roof cover made of PVC-coated polyester fabric, insure the silos are weather resistant (Fig. 7.17).

**Fig. 7.17**   Flexible Silos

## 7.19    Sealing Sheets

Tanks for fluid products, whether it be water or liquid manure, usually have a bottom and resistant side walls composed of juxtaposed prefabricated panels. To better contain fluids and effluents, some of these are lined with special sealing sheets or tank liners. Sealing sheets have proved to be a successful and economical method of solving a large variety of liquid containment problems. They are adaptable to nearly all shapes, sizes and types of tanks, and are resistant to a wide range of chemicals, industrial effluents and other liquids. These urethane blend flexible membranes prevent leakage from under the tank and avoid costly and/or contaminated fluids from escaping into the soil (Fig. 7.18).

**Fig. 7.18**    Sealing Sheets

## 7.20    Flexible Tanks

Flexible tanks are ideal for storing and transporting liquids (drinking water, hydrocarbons, chemical solutions, foodstuffs, industrial or agricultural waste, sludge, etc.). Manufactured using elastomer or plastomer materials, depending on the application, and reinforced with a high strength fabric insert (usually PVC), flexible tanks are a simple and economic solution for several farming applications, thus replacing costly stainless steel tanks, and expensive glass or lead linings (Fig. 7.19).

**Fig. 7.19**   Flexible Tanks

## 7.21    Aquaculture Pond Liners

Liners are used in tanks or ponds, which house fish to provide a controlled environment: one with clear water that can be treated to minimize disease and nuisance weeds. The material must not emit any harmful chemicals. It must have excellent puncture and tear resistance, especially for farming applications where it will have to hold up to harvesting and cleaning, or where the liner is installed over a rocky surface. The liner must be fabricated and installed so that it is watertight-preventing seepage can be a major cost savings. Reinforced polypropylene is often used for these applications (Fig. 7.20).

**Fig. 7.20**   Aquaculture Pond Liners

## 7.22    Agricultural Belting

A conveyor belt consists of two end pulleys, with a continuous loop of material that rotates about them. The pulleys are powered, moving the belt and the product on the belt forward. Conveyor belts are extensively used to transport agricultural materials, such as grain, fodder and various farm equipments. With a narrow window of opportunity when the crop must be harvested, timing is vital in agriculture. As the crops are taken up from the fields and moved through various processing stages, belting plays a key role. Harvesters use rugged PVC conveyor belt with interwoven carcass and cleats of various heights and spacing (Fig. 7.21).

**Fig. 7.21**  Agricultural Belting

## 7.23    Tarpaulins

A tarpaulin, or tarp, is a large sheet of strong, flexible, water-resistant or waterproof material, usually coated with plastic or latex. Tarps have multiple uses, including shelter from the elements (i.e., wind, rain or sunlight), use as a ground sheet for equipment, or covering for protecting vehicles or wood piles. They are also used on outdoor market stalls to provide some protection from the weather. Tarps often have reinforced grommets at the corners and along the sides to form attachment points for rope (Fig. 7.22).

**Fig. 7.22**  Tarpaulins

## 7.24    Anti-Insect Fabric

Anti-insect fabric is a closely woven UV stabilized monofilament polypropylene biomesh used for a variety of applications. It has a strand density of 32 strands per square inch. This provides an average opening between the strands of only half a millimetre. Sheets of biomesh may be used outdoors over a simple framework to effectively guard against pest as small as 0.5 mm. It can also be used in horticultural and agricultural structures covering ventilation openings to block out insects (Fig. 7.23).

**Fig. 7.23**  Anti-Insect Fabric

## 7.25    Flower And Vegetable Support Mesh

High strength polypropylene flower and vegetable support mesh provides crop row stability and assists in preventing stem and bloom damage. Plastic netting can offer less stem abrasion than traditional wire netting and is cost-effective and easily disposed of. Distinctive mesh colour is essential to make the netting highly visible among dense crop foliage for improved handling and reduced accidental cut is shown in Fig. 7.24.

**Fig. 7.24**    Flower and Vegetable Support Mesh

## 7.26 Insulation Nets

Near transparent knitted insulation nets are used in protecting crops from heavy rain, pests and frosts. Their permeable construction allows air and moisture to travel through them at a reduced and controlled rate (Fig. 7.25).

Insulation nets are often used as a horizontal curtain within a structure to create an overhead thermal barrier. They usually provide minimal shading to the crop for improved crop performance in duller and cooler winter/spring months.

**Fig. 7.25** Insulation Nets

## 7.27 Greenhouse Light Reflective Flooring

This flooring cover option is manufactured from a heavy-duty commercial grade of UV-stabilized polypropylene. Its bright white colour provides excellent 'available light' reflectivity from the floor back into dense crop foliage. Greenhouse flooring will resist heavy foot traffic and wear from greenhouse trolleys and is usually highly breathable to limit the likelihood of the covered soil going 'sour' (Fig. 7.26).

**Fig. 7.26** Greenhouse Light Reflective Flooring

## 7.28 Livestock Ground Fabric

High-strength, long-lasting ground fabric is used to keep cattle and other agricultural animals off wet and soft soil patches. Livestock fabrics are used for aggregate roads and paths and under feed lots, as well as for the protection of livestock in stalls or holding areas in farming operations (Fig. 7.27).

**Fig. 7.27** Livestock Ground Fabric

## 7.29 Tree Guards

Tree guards are manufactured from HDPE; they have the ability to form a large range of guards from 60 to 180 cm high and in various diameters. Plastic tree guard mesh offers a versatile method for protecting young trees from animals ranging from rabbits to deer. The tree guards are the best solution for keeping out animals and letting the tree grow naturally (Fig. 7.28).

**Fig. 7.28** Tree Guards

## 7.30    Agro-Protective Garments

During farming, there are chances of harmful pesticides that will penetrate through clothing and will come in contact with the skin and cause a problem. Use of clothing like rubber apron, waterproof outer garment and face mask for longer period is difficult. Hence, selection of clothing is important. These agro-protective garments must be

- lightweight and cheaper,
- protect from pesticides and
- good breathable property and readily washable.

**Fig. 7.29**    Protective Clothing for Pesticide Applicators and Threshing

Traditionally textiles with polyester and cotton with repellent finish were used today which are made from high-tech woven and non-woven with fluorocarbon finish. This new technology of the protective textile not only protects the farmer from harmful pesticides but also keep him comfortable and aesthetically satisfied.

Exposure to pesticides and organic dust is the most important occupational risks among small and marginal farmers. To overcome their occupational health hazard, protective clothing/accessories were designed and tested for their suitability and acceptability. Suitability assessment of designed protective clothing highlighted that all the functional features incorporated in the garments/accessories were assessed to be highly suitable as they provided protection to the wearer without causing any health problem or the hindrance while working. The designed protective clothing for the farm workers reduced their occupational health hazards and increased their work efficiency (Fig. 7.29).

Recommended protective clothing for pesticide applicators is dress jacket with hood and pyjama of water proof fabric (having lining of cotton hosiery fabric), mask resistant to chemicals (respirator particulate), glasses/goggles, sports shoes and surgical gloves. Recommended protective clothing for threshing period is apron with hood and full sleeves with elasticized cuffs (for males), kameez with full sleeves preferably elasticized cuffs (for females), pleated mask/beak mask, glasses/goggles and sports shoes.

# References

1. Subramaniam, V., Poongodi, G.R., Veena Sindhuja, V. Agro-textiles: Production, Properties & Potential. Published on April 2009.

2. World Academy of Science, Engineering and Technology, International Journal of Fashion and Textile Engineering Vol. 2, No. 7, 2015.

3. Hira, M.A. Agro-textile Products & Their Usage. Sasmira, Mumbai.

4. http://www.textilemedia.com/technical-textiles/new-textile-materials/agrotextiles/

5. http://www.textileworld.com/Issues/2005/September/Nonwovens-Technical_Textiles/Agrotextiles-A_Growing_Field

6. http://textilelearner.blogspot.com/2012/02/agro-textiles-general-property.html

7. Shehrawat, P.S. Agro Processing Industries – A Challenging Entrepreneurship For Rural Development. HAU, Hisar, 2006, pp. 8–11.

8. http://www.fibre2fashion.com/industry-article/textile-industry-articles/agro-textiles-a-rising-wave/agro-textiles-a-rising-wave1.asp

9. http://www.indiantextilejournal.com/articles/FAdetails.asp?id=1999

10. http://www.indiantextilejournal.com/articles/FAdetails.asp?id=4709

11. http://www.indiantextilejournal.com/articles/FAdetails.asp?id=884

12. http://www.indiantextilejournal.com/articles/FAdetails.asp?id=4020

13. http://www.technicaltextile.net/articles/agro-textiles/detail.aspx?articleid=4276

14. www.fibroline.com/agrotextiles

15. www.linq.com/agrotex/agrohome

16. www.technical-textiles.net

17. www.magrotexsl.com

18. www.ttf.textiles.org.tw/newss

# 8

# Testing and Evaluation of Agro Textiles

Agriculture forms the backbone of the Indian Economy and one cannot disregard the significant role that agriculture plays in the Indian Economy and in the daily life of its citizens. Yet, food security continues to be a pressing issue in India. In the light of this major challenge, agro textile utilization has helped the agriculture community in attaining increased yield and enhanced quality in agriculture produce. Among its various benefits, agro textiles protect produce from harmful external elements and assist in better soil management.

The primary purpose of textile testing and analysis is to assess textile product performance and to use test results to make predictions about product performance. Product performance must be considered in conjunction with end-use; therefore, tests are performed with the ultimate end-use in mind.

These benefits provide farmers with enhanced productivity and increased yields resulting in further socioeconomic development of the stake holders within the agriculture community. Agro processing sector has experienced expansion during the last five decades, starting with a handful of facilities which were mainly operating at domestic level.

The selection of agro textile product depends on crop needs. Selection of the agro textiles is also greatly influenced by the geographical location. But for any application and supervision, quality control tests are an essential part. Again, proper testing of technical textiles meant for agriculture is critical to ensure their effective performance. The standards evolved for this purpose relate to synthetic agro textiles only and are not uniform.

## 8.1    Importance of Testing

One of the major deterrents for expansion of technical textile market in India is the absence of standards and regulatory legislation. Standards, as a driving factor in the technical textile business, are the framework for any manufacturer of technical textile products. There is a need of standardization and regulation for each product category and its segment, and it will also have a positive impact on the consumption of technical textile product in India.

Standards are the possible way towards ensuring regulatory use of technical textile products. Without having standards in place, the regulatory framework cannot be developed or implemented to the fullest. The standards become more critical from the aspect that several technical textile products have crucial application in infrastructure, live saving applications, personal protection, etc.

## 8.2    Objective of Testing and Analysis

The main objective of testing and analysis are research and development, quality control, comparative testing, analyzing, product future, government regulation, selection of raw materials, product control, process control, process development, product testing, etc. They are elaborated as follows.

### 8.2.1    Research and Development

Textile products are evaluated during the development process. This helps textile scientists determine how to proceed at the each stage of product development. This category also includes testing to study theories of fabric or fibre behaviour. With advances in research and development, new products and processes may require innovative testing procedures that are not provided through standard test methods. Test methodologies developed for a specific research application within one laboratory often gain wider acceptance and eventually are developed into industry-wide standard test methods.

### 8.2.2    Quality Control

Textile products are tested at various stages of production to assure quality processing and products. For example, in dyeing processes, the fabric is evaluated to determine whether it is dyeing evenly. Manufacturers may use quality-control testing as a marketing tool, in that trade names imply to the consumer that certain levels of quality are assumed to be standard for products produced by the manufacturer. Quality-control testing aids the manufacturer in assuring that the expected level of quality is maintained.

### 8.2.3    Comparative Testing

Comparative testing compares two or more products being considered by a company or government agency. For example, a jeans manufacturer may perform a series of tests on denim fabrics from different suppliers prior to deciding which supplier to use for its product. In selecting between competitive products, a fabric manufacturer also may test fibres or yarns from different suppliers.

## 8.2.4    Analyzing Product Failure

Testing is done in this case to pinpoint defects in processing or design. The example of the trouser pockets falls into this category. Results from this type of test can be used to improve products and are also used to determine liability in litigation.

## 8.2.5    Government Regulations

Textile product testing is sometimes performed to meet government regulations. Such regulations may require mandatory testing of products before they can be legally sold. An example of this is the flammability testing of textiles.

    The testing of fibres is generally not so important when dealing with man-made fibres and man-made continuous filament yarns, because they are supplied to customers' requirements and their properties, including length, colour and fineness are determined and controlled during their manufacture.

## 8.3    Standard Test Methods and Specifications

The term standard is used often in regard to testing of products. It may be ambiguous at times, as it can have several different meanings. It can refer to the actual test method or to the minimum acceptable level of performance on a particular test. It is generally used as an adjective, in which we mean 'uniform', 'controlled' or 'widely accepted', as in a standard test method or a standard performance specification. An exception is its use as a proper noun to refer to a specific, numbered, test method or specification.

## 8.3.1    Standard Test Methods

### 8.3.1.1    Development of Standard Test Methods

Test methods are developed for textiles and textile products by several different organizations. They are typically developed in response to a need expressed by an individual manufacturer, a product user or occasionally by a consumer group.

    In most organizations that develop standard test methods, once the test procedure is clearly defined, the proposed method then undergoes inter-laboratory trials. Inter-laboratory testing can reveal problems with procedures that must be corrected, and they can also be used to determine whether the test method is applicable to a particular type of product; for example, does the method work only on woven fabrics or can it also be sued for knits? The primary purpose of the inter-laboratory test is to determine the precision of the test. Precision indicates whether the test will repeatedly produce the same results on the same fabric specimen. Inter-laboratory test determine the reproducibility of the test from one laboratory to another and from one operator to another. A test which has a high level of precision has good inter-laboratory reproducibility and good between-operation reproducibility.

Following inter-laboratory testing and refinement of the method, the proposed test method is submitted to committee vote. When approved by the committee, the method must undergo balloting by other committees. For example, in *American Association of Textile Chemists and Colorists* (AATCC), after a proposed test method on weathering is approved by the weathering research committee, it must be approved by the Editorial Committee, and then by the Technical Committee on Research, which is composed of the chairs from many different research committees. At each level of balloting, input from committee members is sought and is used to improve the test method. At each level, attempts are made to resolve negative votes through written correspondence and conferences.

Finally, once a test method is approved as a standard test method for the organization [such as AATCC or *American Society for Testing and Materials* (ASTM)], the method must undergo periodic reconsideration and re-approval in order to retain as a standard test method. This extensive development and review process are intended to assure that standard test methods meet the needs of users. Test method development and revision are ongoing processes. New test methods are introduced every year, and older methods are dropped in response to the changing needs of the textile, apparel and home furnishing industries and their consumers.

### 8.3.1.2    Format of Standard Test Methods

Test methods usually have a standard form, regardless of which organization developed them. The sections of a test method include the following:

- Test number and name – This usually also includes the year that the method was accepted or revised by the organization.
- Scope and purpose – This states what types of materials are covered by the test method and for what purpose the method was originally intended.
- Definition of terms – Any terms that are not generally understood or that have definitions that are specific to the test method is defined.
- Safety precautions – These are now required for most methods. They prescribe special handling precautions for chemicals or equipment to be used during performance of the test.

- Apparatus and materials – This section describes the instruments, devices and materials that are required to conduct the test.
- Test specimens – The size, number and preparation of test specimens are explained.
- Procedure – This section outlines in detail the steps to follow in performing the test, and any related factors that need to be controlled as the test is performed.
- Evaluation or calculation of results – This explains how to acquire the data or result. It includes explanation of any factors, such as ratings or formulas, needed to determine the results.
- Report – This section indicates what information should be given in the report describing the test results.
- Precision and bias – The precision to be expected from the test is outlined and may know biases in the test are identified.

## 8.3.2    Performance Specifications

Standard performance specifications are based on standard test methods. In other words, the statement of how a fabric must perform in a particular end-use is designated in terms of results on standard tests.

## 8.4    Standard Test Methods

### 8.4.1    Bureau of Indian Standards – India

The Bureau of Indian Standards (BIS) Act (1986) elaborates Indian Standards in relation to any article or process and amends, revises or cancels the standards so established as may be necessary, by a process of consultation involving consumers, manufacturers, government and regulatory bodies, technologists, scientists and testing laboratories through duly constituted committees of the bureau.

For formulation of Indian Standards, BIS functions through a Technical Committee structure comprising of Division Councils, Sectional Committees, Subcommittees and Panels. Division Councils are set up by Standards Advisory Committee in defined areas of industries and technologies for formulation of standards. These include representatives of various interests such as consumers, regulatory and other government bodies, industry, scientists, technologists, testing organizations and consultants. BIS officer is the Member Secretary of the Division Council. The Division Councils set up Sectional Committees within their areas, define their scopes, appoint their Chairmen and members and coordinate their activities.

There are 14 Division Councils and over 650 Technical Committees that have so far developed over 19,000 Indian Standards. Over 350 new and revised standards are being formulated each year by BIS. Indian Standards formulated by BIS cover various aspects such as product standards (specifications), methods of test, codes of practice, guides, recommendations, terminology, dimensions, symbols, etc.

The committee structure of BIS seeks to bring together all those with substantial interest in particular project, so that standards are developed keeping in view national interests and after taking into account all significant view points through a process of consultation. Decisions in BIS technical committees are reached through consensus. As a policy, the standards formulation activity of BIS has been harmonized as far as possible with the relevant guidelines as laid down by the International Organization for Standardization (ISO). BIS, being a signatory to the 'Code of Good Practice' for the preparation, adoption and application of standards (Article 4 of WTO-TBT Agreement, Annex 3), has also accordingly aligned its standards formulation procedure.

## 8.4.2    British Standards – Britain

The standards produced are titled British Standard (BS) XXXX[-P]:YYYY where XXXX is the number of the standard, P is the number of the part of the standard (where the standard is split into multiple parts) and YYYY is the year in which the standard came into effect. BS Institution (BSI) group currently has over 27,000 active standards. Products are commonly specified as meeting a particular British Standard, and in general this can be done without any certification or independent testing. The standard simply provides a shorthand way of claiming that certain specifications are met, while encouraging manufacturers to adhere to a common method for such a specification.

The Kitemark can be used to indicate certification by BSI, but only where a Kitemark scheme has been set up around a particular standard. It is mainly applicable to safety and quality management standards. There is a common misunderstanding that Kitemarks are necessary to prove compliance with any BS, but in general it is neither desirable nor possible that every standard be 'policed' in this way.

### 8.4.2.1    Status of Standards

Standards are continuously reviewed and developed and are periodically allocated one or more of the following status keywords.

- Confirmed – the standard has been reviewed and confirmed as being current.

- Current – the document is the current, most recently published one available.
- Draft for public comment – a national stage in the development of a standard, where wider consultation is sought within the United Kingdom.
- Obsolescent – indicating by amendment that the standard is not recommended for use for new equipment, but needs to be retained to provide for the servicing of equipment that is expected to have a long working life, or due to legislative issues.
- Partially replaced – the standard has been partially replaced by one or more other standards.
- Proposed for confirmation – the standard is being reviewed and it has been proposed that it is confirmed as the current standard.
- Proposed for obsolescence – the standard is being reviewed and it has been proposed that it is made obsolescent.
- Proposed for withdrawal – the standard is being reviewed and it has been proposed that it is withdrawn.
- Revised – the standard has been revised.
- Superseded – the standard has been replaced by one or more other standards.
- Under review – the standard is under review.
- Withdrawn – the document is no longer current and has been withdrawn.
- Work in hand – there is work being undertaken on the standard and there may be a related draft for public comment available.

### 8.4.3    American Society for Testing of Materials – The United States

The ASTM International is an international standards organization that develops and publishes voluntary consensus technical standards for a wide range of materials, products, systems and services. About 12,575 ASTM voluntary consensus standards operate globally. The organization's headquarters is in West Conshohocken, Pennsylvania, about 5 miles (8.0 km) northwest of Philadelphia. ASTM, founded in 1898 as the American Section of the International Association for Testing Materials, predates other standards organizations such as the BSI (1901), International Electrotechnical Commission (1906), Deutsches Institute fur Normung (DIN) (1917), American National Standards Institute (1918), Association Francaise de Normalisation (1926) and ISO (1947).

The ASTM International has no role in requiring or enforcing compliance with its standards. The standards, however, may become mandatory when referenced by an external contract, corporation or government.

- In the United States, ASTM standards have been adopted, by incorporation or by reference, in many federal, state and municipal government regulations. The National Technology Transfer and Advancement Act, passed in 1995, requires the federal government to use privately developed consensus standards whenever possible. The act reflects what had long been recommended as best practice within the federal government.
- Other governments (local and worldwide) also have referenced ASTM standards.
- Corporations doing international business may choose to reference an ASTM standard.
- All toys sold in the United States must meet the safety requirements of ASTM F963, Standard Consumer Safety Specification for Toy Safety, as part of the Consumer Product Safety Improvement Act of 2008. The law makes the ASTM F963 standard a mandatory requirement for toys while the Consumer Product Safety Commission studies the standard's effectiveness and issues final consumer guidelines for toy safety.

### 8.4.4    Deutsches Institute fur Normung – Germany Standards Institute

The DIN (in English, the German Institute for Standardization) is the German national organization for standardization and is the German ISO member body. DIN is a German Registered Association (e.V.) headquarters is in Berlin. There are currently around 30,000 DIN Standards, covering nearly every field of technology.

### 8.5    International Standards for Agro Textiles

Technical textiles are defined as textile materials and products used primarily for their technical performance and functional properties. Unlike conventional textiles where aesthetic value is one of the key usage considerations, technical textiles are used on account of their specific physical and functional properties. Technical textiles are used individually as a stand-alone product, or as a component/part of another product to improve the performance of the product.

Technical textiles are also referred to as industrial textiles, functional textiles, performance textiles, engineering textiles and high-tech textiles. They represent a multi-disciplinary field with numerous end-use applications. Technical textile fabrics have application in almost all major areas of economic activity: aerospace, shipping, sports, agriculture, defence, health care, construction, etc. Non-wovens are the key materials used for manufacturing technical textile products. Non-woven fabrics are engineered fabrics that may be a limited life, single-use fabric or a very durable fabric. Non-woven fabrics provide specific functions such as absorbency, liquid repellence, resilience, stretch, softness, strength, flame retardancy, washability, cushioning, filtering, microbial barrier, sterility, etc. These properties are often combined to create fabrics suited for specific jobs, while achieving a good balance between product use-life and cost. They can mimic the appearance, texture and strength of a woven fabric and can be as bulky as the thickest paddings. In combination with other materials, they provide a spectrum of products with diverse properties and are used alone or as components of apparel, home furnishings, health care, engineering, industrial and consumer goods.

Some of the technical textile products require mandatory prescriptions for their use. The Expert Committee on Technical Textiles constituted by the Ministry of Textiles has also recommended mandatory prescriptions for certain items. One of the major deterrents for the expansion of usage of many technical textile products is the absence of standards and regulatory legislation in India.

In the United States and Western European countries, legislation for mandatory use of such products has led to increase in demand. There is a need of standardization and regulation for each product category and its segment, and it will also have a positive impact on the consumption of technical textile products. Standards are the possible way towards ensuring regulatory use of technical textile products. If regulations pertaining to usage are introduced in Indian context, the full market potential of technical textiles may be realized in an expedited manner, as on one hand, it will create a steady market, whereas on the other hand, the manufacturing sector will be required to upgrade itself to the level of international standards.

Most of the countries of Western Europe (e.g., Belgium, France Germany, Italy, the Netherlands, Switzerland and the United Kingdom) have national standards on the construction, testing and use of various types of synthetic agro textiles. There is already a large volume of trade in agro textiles among the countries of Western Europe but standard procedures for testing different parameters of agro textiles of the producer country.

## 8.6    National Standards for Agro Textiles

There are reportedly 21 BIS for the manufacture, testing, etc., of various types of synthetic agro textiles (Table 8.1). The test parameters normally required for testing agro textiles are given in Table 8.2.

**Table 8.1**    BIS Standard for Testing Agrotech

| S. No. | BIS Standard | Description |
|---|---|---|
| 1. | IS 15351:2008 | Textiles: Laminated high-density polyethylene woven fabric (geomembrane) for water proof lining (First revision) |
| 2. | IS 15907:2010 | Agro textiles: High-density polyethylene woven beds for vermiculture – Specification |
| 3. | IS 4401:2006 | Textiles: Twisted nylon fishnet twines (fifth revision) |
| 4. | IS 4402:2005/ISO 1107:2003 | Textiles – Fishing nets – Netting – Basic terms and definitions (second revision) |
| 5. | IS 4640:1993/ISO 858:1973 | Fishing nets: Designation of netting yarns in the tex system (first revision) |
| 6. | IS 4641:2005/ISO 1530:2003 | Textiles – Fishing nets: Description and designation of knotted netting (second revision) |
| 7. | IS 5815(Part 4):1993/ISO 1805:1973 | Fishing nets: Determination of breaking load and knot breaking load of netting yarns (first revision) |
| 8. | IS 5815(Part 5):2005/ISO 1806:2002 | Textiles – Fishing nets: Determination of mesh breaking force of netting (second revision) |
| 9. | IS 5815(Part 6):1993/ISO 3090:1974 | Netting yarns: Determination of change in length after immersion in water (first revision) |
| 10. | IS 5815(Part 7):1993/ISO 3790:1976 | Fishing nets: Determination of elongation of netting yarns (first revision) |
| 11. | IS 6348:1971 | Basic terms for hanging of netting |
| 12. | IS 6920:1993/ISO 1532:1973 | Fishing nets: Cutting knotted netting to shape (Tapering) |
| 13. | IS 8746:1993/ISO 3660:1976 | Fishing nets: Mounting and joining of netting – Terms and illustrations (first revision) |

| 14. | IS 9945:1999 | Fishing nets: Method for determination of taper ratio and cutting rate (first revision) |
|---|---|---|
| 15. | IS 15788:2008 | Fishing nets: Method of test for determination of mesh size – Opening of mesh |
| 16. | IS 15789:2008 | Fishing nets: Method of test for determination of mesh size – Length of mesh |
| 17. | IS 5508 (Parts 1 to 24) | Guides for fishing gears |
| 18. | IS 7533:2003 | Polyamide monofilament line for fishing |
| 19. | IS 14287:1995 | Polypropylene multi-filament netting twine |
| 20. | IS 6347:2003 | Polyethylene monofilament twine for fishing |
| 21. | IS 16008:2012 | Agro textiles: Shade nets for agriculture and horticulture purposes – Specification (clubbed the specifications of 3 shade net standards, i.e., specifications for shade net 50%, 75% and 90% for agriculture application. Thus, have 1 standard against 4 proposed standards) |

**Table 8.2**    Test Parameters for Testing of Agro textiles

| S. No. | Test Parameters | Test Standard |
|---|---|---|
| 1. | Mass per unit area | ASTM D 5261, ASTM D 3776, IS 1964 |
| 2. | Thickness | ASTM D 1777, IS7702 |
| 3. | Tensile strength (wide width) | ISO 10319 |
| 4. | Tensile strength (strip test) | ASTM D 5035, ISO 13934 Part I |
| 5. | Grab test | ASTM D 5034, IS 1969 |
| 6. | UV resistance (weatherability) | ASTM D 4355, ASTM G 154, IS 7903 |
| 7. | Index puncture resistance test | ASTM D 4833 |
| 8. | California bearing ratio | ASTM D 6241 |
| 9. | Water permeability | ASTM D 4491 |
| 10. | Apparent opening size | ASTM D 4751 |
| 11. | Water vapour permeability | ASTM E 96 |

| 12. | Impact resistance test | ASTM D 1709 |
|---|---|---|
| 13. | Bursting strength test | ASTM D 3786, IS 1966 |
| 14. | Air permeability test | ASTM D 737, IS 11056 |
| 15. | Flame retardancy | IS 11871 |
| 16. | Colour fastness to light | IS 2454 |
| 17. | Water repellency | IS 391 |
| 18. | Seam strength | ASTM D 1683 |
| 19. | Resistance to flexing tester | IS 7016 Part IV |
| 20. | Identification | IS 667 |
| 21. | Resistance to extreme temperatures | Visual observation |
| 22. | Light transmission % | ASTM D 1003 |
| 23. | Wind blockage % | SASMIRA standard method |
| 24. | Durability – soil burial test | ISO 846-1997, AATCC-30-1999 |
| 25. | Antimicrobial test – qualitative | AATCC 100 |
| 26. | Antimicrobial test – quantitative | AATCC-147 |

# References

1. Bureau of Indian Standards. Indian Standards, 271, New Delhi, 1975.

2. Anon. Jute, Kenaf and Allied Fibres – Quarterly Statistics, Food and Agricultural Organisation of The United Nation, 2002.

3. Macmillan, W.G. Indian Textile Journal, Vol. 67, 1957, p. 338.

4. Goswami, B.C., Martindale, I.G., Scardino, F.L. Textile yarns, Technology, Structure and Applications, 1977.

5. Bandyopadhyay, S.B. Frictional properties of jute and some other long vegetable fibers. Part1: General study of characters, Textile Research Journal, Vol. 21, Issue 9, 1951, p. 659.

6. Smith, H.D. Textile Fibres: An Engineering Approach to Their Properties and Utilisation, A S T M Proc, 19th Edgar Marburg Lecture, Vol. 44, 1944, p. 543.

7. https://vibrantgujarat.com/writereaddata/images/pdf/project-profiles/agrotech-textile-application-in-agriculture.pdf

8. http://atira.in/userfiles/file/coe/Compendium/Annex-1%20Agrotech.pdf

9. ficci.in/spdocument/20255/Compendium-Standards-December-2012.pdf

10. www.technotex.gov.in/Compendium%20Standards%202014%20FINAL.pdf

11. www.bis.org.in/sf/pow/txd.pdf

12. https://www.astm.org/Standards/textile-standards.html

# 9
## Market Scenario of Agro Textiles

Agricultural textiles fabrics provide a huge range of woven, non-woven, knitted fabrics that are used for different purposes in forestry, agriculture, horticulture, landscaping, floriculture and aquaculture. Since ages textiles have been used in the field of agriculture for processes in food production and extensively in the fishing industry in the form of products such as nets, ropes and lines. Textiles were also used for the purpose of protecting and covering crops in farmlands.

The revolution of agriculture and use of modern techniques has led the use of textiles for providing varied solutions through weed preventions, soil stabilization, insect barrier and protection, landscaping, soil erosion and frost protection. Agro textiles play an important role in meeting the demand for production of food from limited sources, especially developing countries where the population is rising. Hence the significance of agro textiles is expected to grow in many countries including China, Brazil and India. The market size in India is expected to be worth Rs. 1300 crore for the year 2012–2013.

The global rise in population and growing demand for better and high-quality vegetables and fruits is the main growth driver for agro textiles. According to the United Nation's Food and Agricultural Organization, the global food production capacity will have to increase by 70% from the present level by 2050 to suffice the population. Based on such figures and statistics, the future holds immense potential for agro fabrics and textile products.

Most agro-textile products used in agriculture and horticulture bear features like ability to protect the crop from solar and ultraviolet radiation, microorganisms, water retention and also require being completely biodegradable. Agro textile products are resistant to abrasion, light weighted, flexible and possess high tensile strength making them a viable option to increase the agricultural yields and quality. The most common fibre of choice for agro textiles is synthetic. Polyester, polyethylene (PE), polypropylene (PP), nylon and other synthetic polymers are used in various forms like tape yarn, monofilament and multifilament yarns for the production of bird nets, sunscreen, hail protection, harvesting net and wind shields. Spun-bonded non-wovens for growing plants and PE sheets for mulch mats are also popular uses of textiles in agriculture. Apart from these, natural fibres like wool, jute, coir, sisal, hemp and flax are also used in agro textiles due to their properties of retaining moisture, wet strength and being 100% biodegradable.

Shade cloths are one of the most oldest and known uses of textiles in this field. Different kinds of fabrics woven provide different degrees of shade depending on the specific requirement of the plant. Greenhouses made with light non-woven fabrics are used to create a micro-climate for plants to grow faster and produce a better quality yield. According to the individual need of a plant the temperature, humidity, light, air and even water movements can be controlled.

The PP shades are the products used the most in agro textiles. The resin is treated with additives and pigments which help in providing resistance to the sun rays and weathering. For a high degree of resistance to the sunlight, a black pigmentation is used for agro shades. Extruded nets of the biaxial type of PP, either green or black pigmented, are employed for increased turf grass production. Advanced nets such as these reduce the harvest time, improve handle ability, decrease the irrigation and maintenance cost.

Moreover, agro textiles are also being used for irrigating gardens and landscapes. High-density PE spun-bonded fabrics with micro-pores in the shape of tubes for hydrating plants efficiently are designed. The tubes are either layered besides the plant or buried few inches, under the soil for continuous and consistent watering to the roots of the plant. The many fine fibres in the fabric let the droplets of water to escape from the tube in a uniform manner.

The agro textiles as an industry is so developed that today non-wovens are used as an alternative to plastic films, glass and straw, which were originally used to keep the crops from freezing. Crop covers were initially made up of cotton but now crops and plantations respond well to non-woven row covers protecting the production from an early frost and providing early yields to farmers.

Such innovative use of textiles in agriculture and farming has pushed the field of agro textiles to new limits and into a growing prospective business. Today, the farmers being burdened from the inflating costs of protection are looking for more secure avenues of protecting their agricultural produces. The new generation of farmers are educated and well informed and are switching to more dependable techniques that can help them produce better and high quality yields and crops, yet at the same time reduce costs and maintenance. Agro textile products provide them with just that and much more. The agro-textile industry produces economical and practical solutions to increase crop production, avoid soil from drying and reduce the need for pesticides, fertilizers, water and energy.

The agro-textile market is expected to grow at an average of 10% per annum. In continuation to the consumption of technical textiles worldwide, the importance of agro textiles and its relevance to developing countries like India is well brought out. The current market size for agro textiles is around Rs. 577 crore and could grow to Rs. 900 crore in 2010. The market for agro textiles is expected to grow at 10% per annum, according to a concept paper on agro textiles published by Ficci and Sasmira at the seminar held here on Wednesday.

Worldwide growth of agro textiles does not show a significant growth. The agro-textile market is expected to grow from 1,615,000 tonne (US$6.5 billion) in 2005 to 1,956,000 tonne ($8.1 billion) in 2010. This growth is expected to be at the average rate of 3.9% annually. Agro textile is one of the significant segments of technical textiles products. The market size and potential of the technical textile industry including component in 12 segments have been estimated at Rs. 29,579 crore in 2007–2008. The items covered under agro textiles are fishnets, shade fabrics, mulch mats and woven and non-woven crop covers.

Agro textile is one of the smaller categories of technical textiles, with consumption accounting for around 8.2% by volume and 6.4% by value of the global technical textiles market in 2010. However, this sector is among those with the strongest growth predictions based on the projected increase in global population and the demand for higher quality food. Internationally, the agro□textile market is expected to grow from 1,615,000 tons (US$6.5 billion) in 2005 to 1,958,000 tons (US$8.1 billion) in 2010, at an average growth rate of 3.9% per annum. Developing countries like China, Brazil and India with compound annual growth rate (CAGR) of 7.8% are expected to witness a surge in demand for agro textiles. The global end□use consumption of agro textiles will increase from 3.3% in 2000 to 3.9% by 2010 – according to a David Rigby Associates' study. Unbounded fabrics are increasing in agricultural applications at the expense of woven fabrics. Textile is only a relatively small branch, cost□based reasons the global textile market in agricultural produce about 200 million tons, valued at US$5 billion, an increase of potential. In textiles, agricultural development, production and applications will become an increasing concern.

Agro textile is one of the significant segments of technical textile products, occupying a significant place in terms of volume consumption. Worldwide growth of agro textiles does not show a significant growth. With an improving economy and social scenario in India, a number of enabling factors are expected to positively impact the market for technical textiles. The growth drivers are emerging at both the supply and demand sides and include government support, increased investor interest because of the large untapped market, the way forward to ensure development of the sector lies in promoting international partnerships, training, implementation of policy support, focusing on product innovation, promoting awareness creation programmes and pursuing regulations and standardization of technical textile usage.

Two of the most important sectors where technical textile products have the potential of being used in significant volumes are agro textiles and geo textiles. It is expected that technical textiles will be increasingly used in both these sectors either due to increase in awareness or government regulations. The domestic agro-textiles market is expected to grow at the rate of 8% per annum.

The food security has become a major problem in India and worldwide that is accentuated by the threat of climate change. Some studies have revealed a probability of 10–40% loss in crop production in the country due to the anticipated rise in temperature by 2050–2060. This is the underlying driving force behind the Agrotech sector, to improve and give better yield year after year.

In 2015–2016, technical textiles market reached Rs. 92,499 crore, of which the agro textile share is Rs. 1191 crore, which reflects only 1% of the total. But for a country like India, which thrives on an agricultural economy, the scope is immense.

Besides, jute and coir are also used in some cases. The only problem is about their non-biodegradability, which is coming in the way of promoting them in a big way. But today, recyclable and biodegradable materials are being developed.

Similarly with jute mulch mat use increase in yield of curry leaves was seen to be 64%. Anti-hail/bird nets use has shown increase in average selling price for apples was seen as 67%. Mulch mat use for mango cultivation showed a twofold benefit – an increase of yield of 33% and an increase in average selling price of 30%. Mass awareness and capacity building are important and play a crucial role in the promotion of agro textiles in India via marketing and educational initiatives.

This should involve radio, point-of-sale advertising, information dissemination via Kisan call centres and multi-lingual handbooks. It is also recommended that the benefits of agro textiles are included in the curriculum of graduate courses in agriculture science.

Synthetics have immense scope in textiles, and its potential not fully tapped so far and especially to boost the use of synthetics in cotton blends. SASMIRA has been taking initiatives in Tirupur. This being done in collaboration with the Tirupur Exporters Association and National Institute of Fashion Technology.

In our country, the synthetic textiles are costlier than in China and other countries. A lot of efforts are needed to bring the price and availability on par with the international supply chain. The ratio of synthetics to cotton is 70:30 in developed countries, but in India it is still skewed in favour of cotton at 50:50.

Another field SASMIRA presently engaged in is the use of microorganism for colour dyeing. This is a very novel method and promises chemical-free dyeing of textiles in the future, which will go a long way in sustainability and eco-dyeing. Since this is biotechnology, people involve in biotech has to take it up. Besides, it is commercially viable. This also needs a lot of entrepreneurial talent, and the industry should go for such innovations. About 21 demonstration centres are under construction, and these are expected to give a big boost to agro textile use in various states in the country.

India is emerging as a significant player in technical textiles. The fast-paced economic growth leading to infrastructure creation as well as higher disposable income has made India a key market for the technical textile products. Moreover, the country has developed a foothold in the production of technical textiles owing to its skilled and technical manpower as well as abundant availability of raw material. More investments are underway in this sector; as per the Ministry of Textiles, as on September 2010, 26,163 applications for technical textile projects with a project cost of US$14.5 billion were disbursed under Technology Upgradation Fund Scheme.

Indian technical textile industry is estimated at US$11 billion (2009–2010), with domestic consumption of US$10.3 billion. The industry has witnessed a significant growth of 16% from 2001–2002 to 2009–2010 and is expected to grow at a rate of 11% year-on-year and reach a market size of US$15.1 billion by the year 2012–2013. Domestic consumption is expected to increase to US$14.1 billion by the year 2012–2013 (Fig. 9.1).

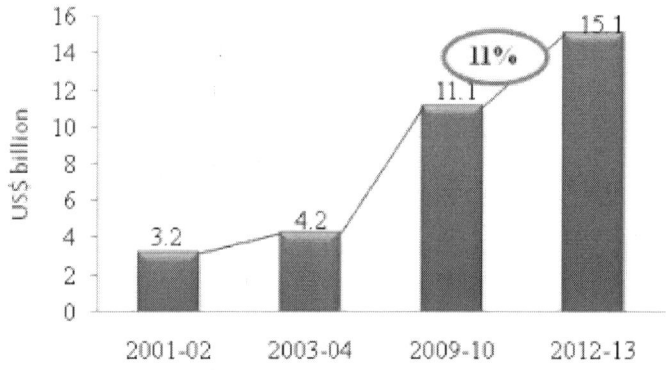

**Fig. 9.1**  Significant Growth of Industry

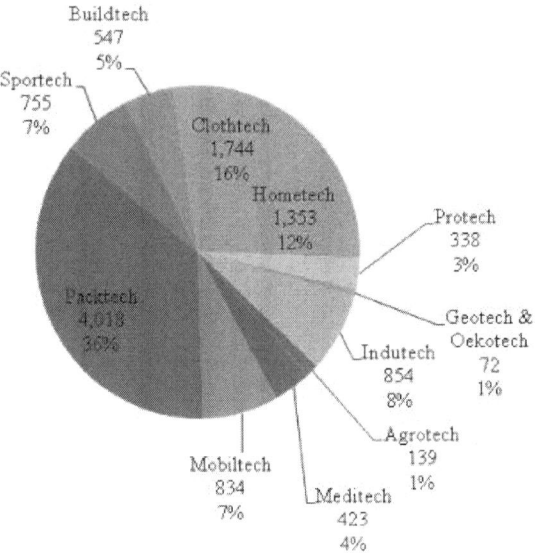

**Fig. 9.2**　Growth of Technical Textile Industry

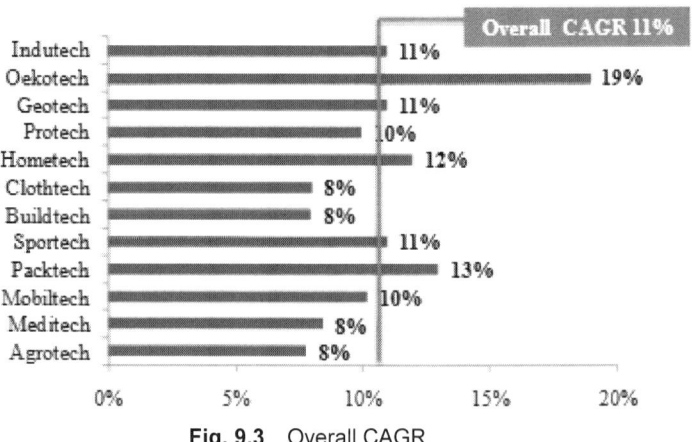

**Fig. 9.3**　Overall CAGR

Agrotech includes technical textile products used in agriculture, horticulture, fisheries and forestry. The technical textile products covered under the segment include shade nets, mulch mats, crop covers, fishing nets, anti-hail nets and bird-protection nets.

Indian Agrotech market is estimated at US$136.4 million in 2009–2010, with domestic consumption of US$119.3 million. As of 2009–2010, fishing nets constituted around 84% of the segment's market, valued at US$114.1 million. This segment is witnessing a significant thrust from the government owing to increasing awareness about the benefits of the usage of shade nets, mulch mats, crop covers, anti-hail nets and bird-protection nets. National Horticulture Mission (NHM) has been actively promoting the usage of Agrotech products (Fig. 9.2).

India through subsidies and annual plans develop the Agrotech products for the states. NHM has included the Agrotech products aimed at plant protection under the protective cultivation in the state-wise action plans. The XI plan envisages to expand the area under mulching by 100,000 hectares. In addition, the XI plan aims at providing assistance for procuring anti-hail nets in the hail prone states like Jammu and Kashmir, Himachal Pradesh and Uttar Pradesh. In line with these developments, the domestic consumption of this segment is expected to increase to around US$153 million by 2012–2013, growing at a CAGR of around 8% over next 3 years. The consumption of mulch mats has the potential to grow at a CAGR of 51% from 2009–2010 to 2012–2013 and that of anti-hail nets and bird-protection nets at 20% (Fig. 9.3). However, the growth of this segment depends upon the performance of agriculture sector and, on the awareness and acceptance of these products by the farming community. The subsidies extended by NHM will continue to play a crucial role in increasing the consumption of these products (Fig. 9.4).

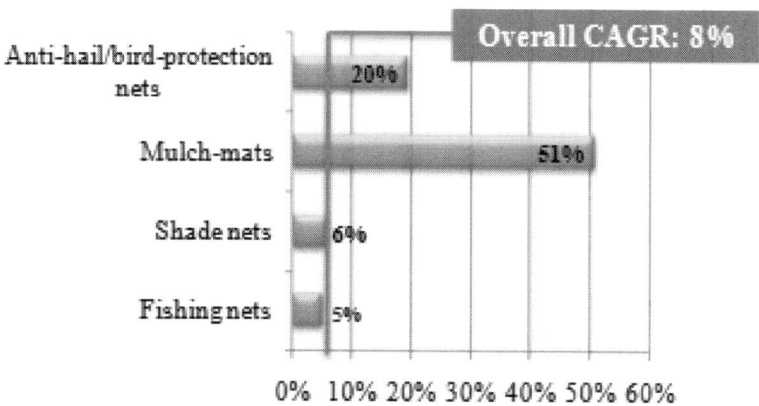

**Fig. 9.4**  Value-wise CAGR for Agrotech Products (from 2009–2010 to 2012–2013)

Majority of the demand for Agrotech products is satisfied by domestic production, with only around 4% of the fishing nets being imported. As far as exports are concerned, crop covers and shade nets are the export-intensive products, as evident from Fig. 9.4. The major export markets include Middle-East countries, United Kingdom, the Netherlands, Italy, Belgium, Poland, New Zealand and Africa.

# References

1. Chaudhary, A., Shahid, N. Technical Textiles in India: The trade perspective. JM International Journal of Management Research. Vol. 2, Issue 6, 2012.

2. Horrock, A.R., Anand, S.C. Handbook of Technical Textiles. Woodhead Publishing & CRC Press, UK, 2004.

3. Ramkumar, S. Technical textiles: A growing necessity for the Indian textile industry. Textile Review, Vol. 6, Issue 1, 2011.

4. Ministry of Textiles. Presentation on Technical Textiles with focus on the use of Agro Textiles, 2006.

5. https://texeducation.wordpress.com/2013/12/30/examples-of-textiles-used-in-agriculture/

6. www.technicaltextile.net/about-technical-textile.aspx

7. www.strategyr.com/MarketResearch/Technical_Textiles_Market_Trends.asp

8. www.marketsandmarkets.com › Top Market Reports

9. https://www.grandviewresearch.com/press-release/global-technical-textiles-market

10. www.sasmira.org/an%20article.pdf

# 10
## Conclusion

Today agro textile plays a significant role to control environment for crop production, eliminate variations in climate, weather change and generate optimum condition for plant growth. Adopting the high-tech farming technique, where textile structures are used, could enhance quality and overall yield of agro products. Textile structures in various forms are used in shade house/ poly house, green house and also in open fields to control environmental factors like temperature, water and humidity. The need of textile goods in the field of agriculture has been stressed and their role in the reduced usage of harmful pesticides and herbicides to render a healthy farming culture underlined. Unique manufacturing techniques and properties of this blend of agro textile sector products whose cost is lesser than that of pesticides and chemical herbicides have been emphasized. 'Agro textiles' gives multidimensional views and solutions to the problems being faced by agro industry. Textiles prove to be flexible in their suitability for specific geographical locations.

Coir is having a very high potentiality in agro textile application. Its moisture retention capability and high wet strength has been excellent and the characteristic has been made use extensively in agro textile applications.

Today, agriculture, horticulture area has realized the need of tomorrow and opting for various technologies to get higher overall yield, quality and tasty agro products. It also avoids agro products get damage from wind, rain and birds. Agro textiles like sunscreen, bird net windshield, mulch mat, hail protection net, harvesting net, etc., can be used for achieving the above goal. Agro textile plays a significant role to help control environment for crop production, eliminate variations in climate, weather change and generate optimum condition for plant growth. Protective screen covering, viz., shade cloth, thermal screen and insect net are tools to further enhance the safety, disease control and productivity of the crop, thus reducing the cost.

'Agro textiles' gives multidimensional views and solutions to the problems being faced by agro industry. Realizing the need of tomorrow, agricultural sector is opting for various technologies to get higher overall yield, quality and tasty agro products.

Agro textile can be used inside greenhouses as well as in the open air, to keep areas safe and tidy. Agro textile improves plant growth and crops in the orchards. Used mainly in planted areas, it provides weed suppression and ground moisture conservation, while allowing roots to breathe and water, air and nutrients to permeate through. This reduces upkeep, maintains higher soil temperatures and promotes more rapid and even plant growth. Many leading landscape architects used for its unrivalled performance, quality and price. Apart from these all application agro textiles are widely used in agriculture, industries, homes and many other areas.

**Agro-net House** – Agro-net house is a framed structure covered by shade nets of different shade factors or any other woven material to allow required sunlight, moisture and air to pass through the gaps. It creates an appropriate microclimate conductive to the plant growth. It is also referred as shade house or net house.

**Agro textiles** – Any textile material used in agriculture, horticulture, forestry, animal husbandry and aquaculture such as fabrics (woven, non-woven or knitted), nets, crop covers, mulch, mats, ropes, etc.

**Agricultural Belting** – A conveyor belt consists of two end pulleys, with a continuous loop of material that rotates about them. The pulleys are powered, moving the belt and the product on the belt forward. Conveyor belts are extensively used to transport agricultural materials, such as grain, fodder and various farm equipments. With a narrow window of opportunity when the crop must be harvested, timing is vital in agriculture. As the crops are taken up from the fields and moved through various processing stages, belting plays a key role. Harvesters use rugged polyvinyl chloride (PVC) conveyor belt with interwoven carcass and cleats of various heights and spacings.

**Air Laid Non-woven** – An air laid web that has been bonded (see Bonding) by one or more techniques to provide fabric integrity.

**Anti-Insect Fabric** – Anti-insect fabric is a closely woven ultraviolet (UV)-stabilized monofilament polypropylene (PP) biomesh used for a variety of applications. It has a strand density of 32 strands per square inch. This provides an average opening between the strands of only half a millimetre. Sheets of biomesh may be used outdoors over a simple framework to effectively guard against pest as small as 0.5 mm. It can also be used in horticultural and agricultural structures covering ventilation openings to block out insects.

**Aquaculture Liners** – Liners are used in tanks or ponds, which house fish to provide a controlled environment: one with clear water that can be treated to minimize disease and nuisance weeds. The material must not emit any harmful chemicals. It must have excellent puncture and tear resistance, especially for farming applications where it will have to hold up to harvesting and cleaning, or where the liner is installed over a rocky surface. The liner must be fabricated and installed so that it is watertight-preventing seepage can be a major cost savings. Reinforced PP is often used for these applications.

**Bird Protection Nets** – UV-stabilized nets made from polyester or light weight PP monofilaments, polyethylene tape yarns and sometimes knitted nylon nets, used to protect crops against damage caused by aerial bird raids and attacks from larger ground animals such as pigeons, butterflies and other similar birds.

**Braided Fishing Line** – One of the of earliest types of fishing line, made from natural and synthetic fibres is known for its high knot strength, lack of stretch and great overall power in relation to its diameter. Braided fishing lines tend to have good resistance to abrasion. Their actual breaking strength will commonly well exceed their pound-test rating.

**Cherry Covers** – Cherry covers are PP nets used for protection of cherry crop throughout the season from frost, rain, hail and wind.

**Drainage Textiles** – Drainage textiles are instrumental in solving the problems created by surface water and poorly drained soils. Geotextiles are mostly useful in drainage management for their filtration capabilities. Simply put, the geotextile, which can consist of a woven or a non-woven fabric, retains the soil while water passes through the fabric and into the drainage collection system.

**Fishing Lines** – Ropes and lines made of fibre lengths, twisted or braided together to provide tensile strength. They are used for pulling, but not for pushing. The availability of reliable and durable ropes and lines has had many consequences for the development and utility of fishing nets, and influences particularly the scale at which the nets can be deployed.

**Fishing Net** – A net used for fishing and consists of usually meshes formed by knotting a relatively thin thread. Modern nets are usually made from monofilament or multifilament of nylon, polyester or high-density polyethylene (HDPE), although nets of organic polyamides such as wool or silk thread are also used.

**Flexible Silos** – Natural products such as grain, animal feed and food place very special demands with regard to conservation on the farm. Flexible textile silos, made of active-breathing, dust-tight, very strong and durable polyester fabric, guarantee constant grain quality and healthy livestock. Contrary to solid silos, flexible silos will also limit condensation, which means that the formation of mould is effectively prevented. Finally, the galvanized steel frame and the jacket and roof cover made of PVC-coated polyester fabric, insure the silos are weather resistant.

**Flexible Tanks** – Flexible tanks are ideal for storing and transporting liquids (drinking water, hydrocarbons, chemical solutions, foodstuffs, industrial or agricultural waste, sludge, etc.). Manufactured using elastomer or plastomer materials, depending on the application and reinforced with a high-strength fabric insert (usually PVC), flexible tanks are a simple and economic solution for several farming applications, thus replacing costly stainless steel tanks, and expensive glass or lead linings.

**Flower and Vegetable Support Mesh** – High-strength PP flower and vegetable support mesh provides crop row stability and assists in preventing stem and bloom damage. Plastic netting can offer less stem abrasion than traditional wire netting and is cost-effective and easily disposed of. Distinctive mesh colour is essential to make the netting highly visible among dense crop foliage for improved handling and reduced accidental cuts.

**Frost Cover Fabrics** – Different climates require varying levels of protection. Used primarily by professional growers for the protection of plants and golf course greens and tee boxes, frost cover fabrics help protect vegetation in wintertime, or from sudden drops in temperature during unseasonable weather, which can cause extensive damage to landscaping and crops.

**Garden Fencing Mesh** – A heavy duty UV-stabilized polyethylene barrier fencing mesh for many applications in horticulture and gardening such as garden fence, or keeping pets like dogs confined or as windbreak fencing. It can be used as a snow fence and sand fencing in extreme conditions.

**Greenhouse** – A framed or inflated structure covered with transparent or translucent materials and large enough to walk in and grow crops under partial or fully controlled environmental conditions.

**Grass Protection Matting** – A reusable heavy duty HDPE reinforcement mesh bonded to a strong needle-punched non-woven fabric used to provide two functions for the temporary protection and access requirements on grass surfaces depending whether the mesh or fabric is laid upwards.

**Greenhouse Light Reflective Flooring** – This bright white flooring cover manufactured from a heavy-duty commercial grade of UV-stabilized PP provides excellent 'available light' reflectivity from the floor back into dense crop foliage. Greenhouse flooring will resist heavy foot traffic and wear from greenhouse trolleys and is usually highly breathable to limit the likelihood of the covered soil going 'sour'.

**Ground Cover Fabric** – The term 'ground cover' refers to any plant material low to the ground, made from PP/polyester that can be used to cover areas where grass will not grow or is not desired. Ground cover fabrics are made up of natural or synthetic materials that are either woven or non-woven products, spun bonded or needle punched used to suppress weed growth around the crop by covering the soil, blocking extreme climatic conditions of sunlight or cold.

**Hails Protection Nets** – UV stabilized polyethylene monofilament nets in woven/knitted form used to cover plants and fruit orchards usually of small mesh size to protect them from being damaged by hailstones, but does not restrict their growth.

**Harvesting Nets** – Nets usually made from polyethylene or polyester monofilaments or polyethylene tape yarns used for collecting fruits which fall off the tree when they are ripe, or which have to be shaken off the tree.

**Horticulture** – The science or art of cultivating fruits, vegetables, flowers or ornamental plants, medicinal plants and their processing and preservation. Etymologically, 'horticulture' can be broken down into two Latin words: hortus (garden) and colere (cultivate).

**Insect Protection Nets** – A light hand net made of fine woven screens generally made of synthetics fibres such as nylon, polyester, etc., used to protect plants from pests attack or for catching insects.

**Insulation Nets** – Near transparent knitted insulation nets are used in protecting crops from heavy rain, pests and frosts. Their permeable construction allows air and moisture to travel through them at a reduced and controlled rate. Insulation nets are often used as a horizontal curtain within a structure to create an overhead thermal barrier. They usually provide minimal shading to the crop for improved crop performance in duller and cooler winter/spring months.

**Landscape Fabric** – Landscape fabric is used for weed control, a central element in achieving low-maintenance landscaping. Effective weed control means a reduction in actual weeding or in herbicide use – both unsavory landscaping tasks. Thermally spun-bonded fabrics are said to be more effective than woven or needle-punched geotextiles in preventing fine roots and rhizomes from penetrating the fabric. While woven fabrics are very strong, they offer many spaces for weeds to penetrate. Needle-punched fabrics have loose threads of material that plants can easily grow through. As for thermally spun-bonded fabrics, these have fibres fixed in place, keeping roots from penetrating.

**Leno Bag** – A bag made from leno fabric (e.g., PP, HDPE, etc.) generally used for packaging and storage of fruits and vegetables.

**Leno Fabric** – A fabric in which warp threads are made to cross one another, between the picks, during leno weaving. The crossing of the warp threads may be a general feature of plain leno fabrics (as in marquisette and some gauzes and muslins) or may be used in combination with other weave (as in some cellular fabrics). Leno weave is also called as gauze or doup weave.

**Livestock Ground Fabric** – High-strength, long-lasting ground fabric used to keep cattle and other agricultural animals off wet and soft soil patches. Livestock fabrics are used for aggregate roads and paths and under feed lots, as well as for the protection of livestock in stalls or holding areas in farming operations.

**Milk Filters** – Non-woven fabrics are used to filter the milk in automatic milking systems.

**Mulch** – Mulch is an aesthetic ground cover applied to planted areas at the time of planting to suppress weed growth. It covers the soil, blocks light and prevents the competitive weed growth around seed links. Mulch is not intended to replace vegetation as an exclusive ground cover, nor is it to be utilized as a growing medium in the landscape.

**Mulch Mats** – A mat used to cover soil around a plant root zone with a view to modify soil environment favourably to conserve soil moisture, moderate soil temperature and suppress weed growth in horticultural crop production.

**Non-woven Fabric** – A fabric made directly from a web of fibre, without the yarn preparation necessary for weaving and knitting.

**Plant Net** – Vertical or tiered net made from polyethylene or PP monofilaments to keep the fruits, which grow close to the ground, away from the damp soil by allowing them to grow through vertical or tiered nets to keep the amount of decayed fruit to a minimum.

**Poultry Curtains** – Used more and more instead of traditional chicken wire to protect precious flocks of farm fowl, poultry curtains offer light control, thermal protection and ventilation control, even at sub-freezing temperatures.

**Root Ball Net** – Nets generally made of synthetic fibres used for protection of young plants so that root system is not damaged when they are dug up, transplanted or replanted.

**Sealing Sheets** – Tanks for fluid products, whether it be water or liquid manure, usually have a bottom and resistant side walls composed of juxtaposed prefabricated panels. To better contain fluids and effluents, some of these are lined with special sealing sheets or tank liners. Sealing sheets have proved to be a successful and economical method of solving a large variety of liquid containment problems. They are adaptable to nearly all shapes, sizes and types of tanks and are resistant to a wide range of chemicals, industrial effluents and other liquids. These urethane blend flexible membranes prevent leakage from under the tank and avoid costly and/or contaminated fluids from escaping into the soil.

**Shade-net House** – A framed structure made up of either galvanized iron (GI) pipe or mild steel angle, or wood or bamboo or GI wire or RCC pole and covered with UV-stabilized polyethylene monofilaments in woven and knitted form and is used in saving food grains, plants from sun, UV rays. Shade nets have different shading area percentages depending upon the type of plant grown to permit controlled sunlight sufficient for growth of plant by creating partially controlled climatic conditions of light intensity and heat during the day time to provide plants a comfortable stress-free growth.

**Shrimp Push Nets** – A triangular shape hand net with wooden or metal frame used to collect shrimp. A handle is attached to the frame and pushed along the surface of sand to collect the shrimp.

**Shade House** – A greenhouse style structure, covered in porous shade cloth for the sole purpose of provided shade for the contents.

**Shade Nets** – Generally made of UV-stabilized PP or HDPE in knitted and woven form used for growth of plants in horticulture, floriculture and sericulture.

**Shade Cloth** – Keeping the sun from burning vegetation is critical in farming. Shade cloths, most often manufactured from a commercial grade of UV-stabilized yarn, offer the protection needed for more sensitive plants and for several other cultures in very hot regions or times of the year. Specific applications of shade cloths include shading ginseng, tropical ornamental plants, ferns, crops of all sorts and even livestock.

**Sun Protection Nets** – Warp-knitted nets made from UV resistant synthetic fibres used in hot countries for protecting plants and people from harmful UV rays and heat by placing over play areas, car parks, patios and terraces.

**Tarpaulins** – A tarpaulin, or tarp, is a large sheet of strong, flexible, water resistant or waterproof material, usually coated with plastic or latex. Tarps have multiple uses, including shelter from the elements (i.e., wind, rain or sunlight), use as a ground sheet for equipment or covering for protecting vehicles or wood piles. They are also used on outdoor market stalls to provide some protection from the weather. Tarps often have reinforced grommets at the corners and along the sides to form attachment points for rope.

**Textile Irrigation Systems** – Optimal moisture management can be achieved using specialized soil covering materials. These fabrics, made from a multilayer high-performance textile, retain soil moisture, such as mulch, and insure a most favourable contact between precious water resources and the plant's roots, thus improving critical aspects of greenhouse tree nursery production. Textile irrigation systems currently being deployed generally have one layer that acts as a reservoir from which water is distributed equally and continuously throughout its surface, as well as a layer made from a light, low-density textile that prevents evaporation while transporting water to the pots through capillarity.

**Turf Protection Net** – A protective soil cover of straw, wood, coconut fibre or other suitable plant residue, or plastic fibres formed into a mat, usually with a plastic or biodegradable mesh on one or both sides. Erosion mats are rolled products available in many varieties and combination of materials and with varying life spans.

**Wind Protection Fabric** – Knitted windshield constructed from a commercial grade of UV-stabilized yarn is often used to protect crops and structures from wind damage. Air exhaust deflection fabrics are otherwise most useful in odour control. These are often seen in swine and dairy house production. Wind protection fabric is available in a variety of aesthetically pleasing colours so structures would not by an eyesore on the farm.

# Index